Practical Spreadsheet Modeling Using @Risk

Practical Spreadsheet Modeling Using @Risk

Dale E. Lehman
Huybert Groenendaal

CRC Press
Taylor & Francis Group
Boca Raton London New York

CRC Press is an imprint of the
Taylor & Francis Group, an **informa** business

A CHAPMAN & HALL BOOK

eResource: The link for students to download the software that accompanies this book is: https://www.palisade.com/bookdownloads/taylorfrancis.asp

CRC Press
Taylor & Francis Group
6000 Broken Sound Parkway NW, Suite 300
Boca Raton, FL 33487-2742

© 2020 by Taylor & Francis Group, LLC
CRC Press is an imprint of Taylor & Francis Group, an Informa business

No claim to original U.S. Government works

Printed on acid-free paper

International Standard Book Number-13: 978-0-367-17386-9 (Hardback)

This book contains information obtained from authentic and highly regarded sources. Reasonable efforts have been made to publish reliable data and information, but the author and publisher cannot assume responsibility for the validity of all materials or the consequences of their use. The authors and publishers have attempted to trace the copyright holders of all material reproduced in this publication and apologize to copyright holders if permission to publish in this form has not been obtained. If any copyright material has not been acknowledged, please write and let us know so we may rectify in any future reprint.

Except as permitted under U.S. Copyright Law, no part of this book may be reprinted, reproduced, transmitted, or utilized in any form by any electronic, mechanical, or other means, now known or hereafter invented, including photocopying, microfilming, and recording, or in any information storage or retrieval system, without written permission from the publishers.

For permission to photocopy or use material electronically from this work, please access www.copyright.com (http://www.copyright.com/) or contact the Copyright Clearance Center, Inc. (CCC), 222 Rosewood Drive, Danvers, MA 01923, 978-750-8400. CCC is a not-for-profit organization that provides licenses and registration for a variety of users. For organizations that have been granted a photocopy license by the CCC, a separate system of payment has been arranged.

Trademark Notice: Product or corporate names may be trademarks or registered trademarks, and are used only for identification and explanation without intent to infringe.

Visit the Taylor & Francis Web site at
http://www.taylorandfrancis.com

and the CRC Press Web site at
http://www.crcpress.com

For my wife Nancy and son Jesse

(DEL)

For my wife Rebecca and daughters Lilian and Violet

(HG)

Contents

Preface xi
Acknowledgments xiii
Authors xv
Introduction xvii

1. Conceptual Maps and Models 1
    1.1   Introductory Case MoviePass 1
    1.2   First Steps: Visualization 2
    1.3   Retirement Planning Example 7
    1.4   Good Practices with Spreadsheet Model Construction 12
    1.5   Errors in Spreadsheet Modeling 12
    1.6   Decision Analysis 14
    1.7   Conclusion: Best Practices 15
    Exercises 16
    Notes 21

2. Basic Monte Carlo Simulation in Spreadsheets 23
    2.1   Introductory Case: Retirement Planning 23
    2.2   Risk and Uncertainty 23
    2.3   Scenario Manager 25
    2.4   Monte Carlo Simulation 27
            2.4.1   Generating Random Numbers 27
            2.4.2   Monte Carlo Simulation for MoviePass 28
    2.5   Monte Carlo Simulation Using @Risk 28
    2.6   Monte Carlo Simulation for Retirement Planning 38
    2.7   Discrete Event Simulation 42

Exercises 44
Notes 46

3. Selecting Distributions 49
    3.1   First Introductory Case: Valuation of a Public Company Using Expert Opinion 49
    3.2   Modeling Expert Opinion in the Valuation Model 50
    3.3   Second Introductory Case: Value at Risk—Fitting Distributions to Data 55
    3.4   Distribution Fitting for Value at Risk—Parameter and Model Uncertainty 56
            3.4.1   Parameter Uncertainty (More Advanced, Optional) 59
            3.4.2   Model Uncertainty (Most Advanced, Optional) 65
    3.5   Third Introductory Case: Failure Distributions 65
    3.6   Commonly Used Discrete Distributions 67
    3.7   Commonly Used Continuous Distributions 71
    3.8   A Brief Decision Guide for Selecting Distributions 73
    Exercises 74
    Notes 78

4. Modeling Relationships 81
    4.1   First Example: Drug Development 81
    4.2   Second Example: Collateralized Debt Obligations 84
    4.3   Multiple Correlations Example: Cockpit Failures 88
    4.4   Copulas Example: How Correlated Are Home Prices? 92
    4.5   Empirical Copulas 97
    4.6   Fifth Example: Advertising Effectiveness 100
    4.7   Regression Modeling 101
    4.8   Simulation within Regression Models 105
    4.9   Multiple Linear Regression Models 107
    4.10   The Envelope Method 111
    4.11   Summary 113
    Exercises 114
    Notes 118

5. Time Series Models 121
   5.1 The Need for Time Series Analysis: A Tale of Two Series 121
   5.2 Introductory Case: Air Travel and September 11 125
   5.3 Analyzing the Air Traffic Data and 9/11 127
   5.4 Second Example: Stock Prices 130
   5.5 Types of Time Series Models 132
   5.6 Third Example: Soybean Prices 133
   5.7 Fourth Example: Home Prices and Multivariate Time Series 134
   Exercises 137
   Notes 141

6. Additional Useful Techniques 143
   6.1 Advanced Sensitivity Analysis 143
   6.2 Stress Testing 145
   6.3 Non-Parametric Distributions 147
   6.4 Case: An Insurance Problem 151
   6.5 Frequency and Severity 152
   6.6 The Compound Distribution 160
   6.7 Uncertainty and Variability 161
   6.8 Bayesian Analysis 163
   Exercises 169
   Notes 173

7. Optimization and Decision Making 177
   7.1 Introductory Case: Airline Seat Pricing 177
   7.2 A Simulation Model of the Airline Pricing Problem 177
   7.3 A Simulation Table to Explore Pricing Strategies 179
   7.4 A Stochastic Optimization Solution to the Airline Pricing Problem 181
   7.5 Optimization with Multiple Decision Variables 187
   7.6 Adding Constraints 190
   7.7 Efficient Frontier 191

7.8	Stochastic Dominance	196
7.9	Summary	200
	Exercises	201
	Notes	206

Appendix: Risk Analysis in Projects 209

Index 217

Preface

This book is about building spreadsheet models to support decisions that involve risk. While there are a number of excellent books on this subject, we felt that there is a need for a book that (1) incorporates some of the latest techniques in risk modeling; (2) is aimed at practitioners and applications; and (3) uses an appropriate pedagogy, including real data and meaningful examples. Risk analytics is a developing field, and we have striven to present examples that illustrate the power of numerous techniques, including methods that are not (yet) very common. Our experience with people who need to use these tools "in the field" is that they need material that is theoretically sound but also practical and straightforward enough for them to use in their work. To this end, we have deferred the focus of the underlying mathematics and concentrated on demonstrating how the techniques and methods can be used correctly within a spreadsheet-based environment to help make decisions under risk.

Who This Book Is For

This book is written for anyone interested in conducting a risk analysis to support a decision. This applies to analysis in business, engineering, environmental planning, public policy, health care, or many other fields that can benefit from modeling in spreadsheets. If you intend to use spreadsheets for the purposes of decision support, rather than merely a placeholder for numbers, then it is likely appropriate to better understand the risk in your decision by conducting a simulation. Consequently, this book may appeal to business students, students using quantitative analysis in other fields, or applied practitioners.

This book is written at a beginner to intermediate level appropriate for graduate students or advanced undergraduates interested in simulation modeling. We have kept the mathematics to a minimum—more technical descriptions of many topics are available from several reference sources, but minimal mathematical background is required in order to use this book. We do assume that the reader is familiar with how to use Microsoft Excel and understands basic statistical concepts (such as mean, standard deviations, percentiles, and confidence intervals). Readers without this background will probably, from time to time, need to supplement this text with additional materials or a web search.

This book is suitable for use in courses ranging from a few days to an entire semester. For a short course, some of the material can be preassigned, and the end-of-chapter problems can be used to practice the techniques covered within the chapter. For longer courses, the entire set of end-of-chapter problems can be covered, which spans a range of industries from health to finance (as explained in the next section). We have also found it to be valuable to add a project-based component after covering the material in the book within a semester-long class.

This book uses the @Risk software throughout and comes with a one-year trial license to the software. Specifically, the analyses and figures presented in the book use @Risk version 7.6. New software releases may have differing appearances and additional functionality, but are not expected to vary much from what is presented here. However, the focus of the book is not on software, but on the different techniques and methods in simulation that can be used to build accurate and useful models to support decision making. We regularly use many different software packages within our work, each with advantages and disadvantages. Users of spreadsheet-based Monte Carlo simulation software products other than @Risk will find that much of the material applies equally well to them.

What Additional Features Supplement This Book

Several features are available to supplement the text. The book website (http://www.EpiXAnalytics.com/LehmanGroenendaal.html) has links to the following:

- The spreadsheet models used in the text
- Text errata
- ModelAssist, a free online resource that contains 150+ @Risk models and describes 500+ risk analysis–related subjects

In addition, for qualified instructors, a teaching guide and exercise solutions are available from CRC Press/Taylor & Francis.

Acknowledgments

My students have always been a source of inspiration, useful criticism, and helpful guidance for me. In particular, Theodore Anders, Joshua Boots, Andrew Kluesner, Michael Griebel, Chad Kruse, Chris Sindahl, Kimberly Sindahl, and Andrew Will were invaluable as guinea pigs and collaborators for this book. My sincere thanks to Huybert, the best risk analyst I know, for agreeing to participate in this endeavor. My gratitude is also due to Greg Nolder, who assisted greatly with a previous book on risk modeling that has served as a basis for the current book. Thanks to the staff at Palisade Corporation for developing the excellent @Risk software and providing the license to accompany this book. My greatest thanks are for my wife, Nancy, who tolerated all the time we could have been doing something else, and who put up with my endless distractions with writing.

Dale E. Lehman

Thanks to my colleagues and peers in the field of risk analysis and risk modeling over the years, from whom I've learned much. Dale, thanks for being such great coauthor, and for keeping the book very practical and hands on. First, I am grateful to all our consulting and training clients with whom I've had the privilege to work over the years. There are few things as professionally satisfying as collaborating with interesting and diverse people on complex and challenging real-life problems.

Huybert Groenendaal

Authors

Dale E. Lehman, PhD, is Professor of Business Administration and Director of the EMBA in Business Analytics at Loras College. He has held positions at a dozen universities in the United States, Europe, and Asia, and has an extensive consulting practice, primarily in the telecommunications and information industries. He has served as an expert witness in numerous state and federal regulatory proceedings. He holds a BA in economics from SUNY at Stony Brook and MA and PhD degrees from the University of Rochester (New York). He has authored numerous articles and three books on topics related to microeconomic theory, decision making under uncertainty, and public policy, particularly concerning telecommunications and natural resources.

Huybert Groenendaal, PhD, is a Managing Director at EpiX Analytics. As a consultant, he helps clients using risk analysis modeling techniques in a broad range of industries. He has extensive experience in risk modeling in business development, financial valuation, and R&D portfolio evaluation within the pharmaceutical and medical device industries, but also works regularly in a variety of other fields, including investment management, health and epidemiology, and inventory management. He also teaches a number of risk analysis training classes, gives guest lectures at a number of universities, and is adjunct professor at Colorado State University. He holds an MSc and PhD from Wageningen University and an MBA in finance from the Wharton School of Business.

Introduction

Risk: What This Book Is About

Financial meltdown, oil spills, climate change: we live in a risky world. Alan Greenspan (2007) has called it "The Age of Turbulence."[1] A search of the Business Source Premier database for the subject terms "risk" or "uncertainty" yields the following number of publications over each of the past six decades:

- 1960s: 1828 publications
- 1970s: 6228 publications
- 1980s: 15,122 publications
- 1990s: 76,078 publications
- 2000s: 212,222 publications
- 2010s: 294,953 publications

The marked increase in attention is evident. This does not necessarily mean the world has become a riskier place. Indeed, in a very real sense, risk has always been part of the human experience. Arguably, risks for early humans were greater—predators, natural catastrophes, and disease were more threatening and severe than they are today. So, we probably do not live in a uniquely risky age. But we do live in the age of *risk analysis*. Never before have so many people had access to the software tools to conduct sophisticated (as well as simple) risk analyses.[2]

Primitive humans did not conduct sophisticated risk analyses. They mostly relied on instinct.[3] Modern risks, the products of complex technologies and economies, challenge us, and our instincts are often maladapted to making good decisions. In particular, human frailties when dealing with risky situations and probabilities have been well documented.[4] Humans developed a variety of heuristics for decision making in the face of these uncertainties—heuristics that can often contain cognitive biases.[5] As Stephen Few puts it "[C]ognitive biases are not a sign of stupidity but remnants of behaviors that worked admirably in our ancestral environment. That environment has now largely vanished, however, and the biases that we developed to survive in it do not always serve us well today."[6]

Decision making when faced with complex modern risks cannot be left to instinct alone: a more systematic approach can provide great insight and better understanding to improve decision making. At the same time, spreadsheets are ubiquitous and used in business, engineering, policy analysis, and virtually anywhere people conduct analysis to support decisions. This book is about the marriage of these two evolutions: risk analysis and spreadsheet modeling. Our belief is that a number of tools are readily available that permit sophisticated risk analysis in a variety of situations that can be conducted without needing an extensive mathematical background.

As data exponentially expands in availability, and as algorithms increasingly automate many analysis tasks, it is our belief that the most important decisions involving risk still require human judgment. Indeed, many management decisions can be automated—it is precisely the decisions where uncertainty looms large that should not be left to algorithms alone. It is these decisions—where human judgment is required, but data and algorithms can help—for which simulations become a useful decision-making tool. That is the subject of this book.

How This Book Is Organized and How to Use It

The first three chapters comprise material covering how to construct spreadsheet models, how to build meaningful spreadsheet models, how to integrate simulation modeling into spreadsheets, and how to choose distributions to represent key uncertainties. Advanced readers can self-study the first two chapters, but all readers should read these three chapters. The remaining Chapters (4–7) need not be covered in sequence, nor in entirety. Chapter 4 focuses on estimating relationships between uncertain variables, and on using simulation to represent the uncertainty about these relationships. Chapter 5 deals exclusively with time series data and forecasting. Chapter 6 contains a number of further topics for risk analysis, including frequency-severity modeling and Bayesian modeling. Chapter 7 examines optimization under uncertainty.

Also, there is an appendix that describes how to integrate simulation with project management, utilizing @Risk and Microsoft Project.

Each chapter has eight end-of-chapter problems. They span a range of industries, including

- Health care
- Transportation
- Finance and insurance
- Consumer/retail
- Technology
- Natural resources
- Manufacturing
- Sports and entertainment

This forms a matrix of seven chapters by eight sectors. Readers interested in a particular area can focus on the chapter problem devoted to that sector. We believe that risk analysis requires repeated application to novel situations, so these problems are essential to learning how to conduct practical risk analysis. We also believe the diversity of these examples will help develop modeling skills for all of these sectors, so we encourage readers to utilize as many of these problems as time permits.

Before attempting the end-of-chapter problems, readers should verify that they comprehend the material in each chapter. The best way to do this is to reproduce the analyses shown in the text, using the spreadsheet models from the book website.[7]

Notes

1. Greenspan, A. 2007. *The Age of Turbulence*, Penguin Press.
2. For an excellent history of risk analysis, see Bernstein, P. L. 1998. *Against the Gods: The Remarkable Story of Risk*. John Wiley & Sons.
3. Malcolm Gladwell (Blink: *The Power of Thinking without Thinking*, Back Bay Books, 2007) describes our abilities to quickly react under uncertainty. However, even where our instantaneous reactions serve us well, there are usually years of preparation, education, and training that permit our instincts to perform well at these times.
4. The classic reference is Kahneman, D., Slovic, P. and Tversky, A. 1982. *Judgment under Uncertainty: Heuristics and Biases*. Cambridge University Press. A more comprehensive and recent reference source is Koehler, D. J. and Harvey, N. editors, 2004. *Blackwell Handbook of Judgment and Decision Making*. Wiley-Blackwell.
5. The heuristics have been characterized as System 1 thinking, to be distinguished from the effortful logical analysis of System 2. See Kahneman, D. 2011. *Thinking Fast and Slow* (2011, Farrar, Strauss and Giroux) for an exposition of this approach.
6. Stephen, F. 2019. *The Data Loom: Weaving Understanding by Thinking Critically and Scientifically with Data*. Analytics Press, 18.
7. In theory, all simulation results in the book can be reproduced precisely by running 10,000 simulations with a the seed value set equal to 1. It is possible that use of a different version of Excel or @Risk, or slight differences in the layout of a spreadsheet (in case that you make any changes to the models) may cause the results to diverge, but the differences should be minor.

1
Conceptual Maps and Models

Learning Objectives
- Use visualizations of a situation to sketch out the logic for a model.
- Translate the visualization into a spreadsheet model.
- Develop good procedures for building and documenting spreadsheet models.
- Appreciate the prevalence of errors in spreadsheets and learn how to prevent them.

1.1 Introductory Case MoviePass

MoviePass is (as of January 2019)[1] a subscription-based service offering discounted pricing for seeing movies in theaters. It was founded in 2011, initially charging $50 per month. It has gone through several pricing iterations, at one point offering unlimited admission to one movie per day, for $9.95 per month. That pricing caused a surge in subscriptions: from 12,000 in August 2017 to more than 3 million in June 2018.[2] The wildly successful growth in subscribers was not matched by business success, however. Losses for 2017 were reported as $150.8 million, up from $7.4 million in 2016. While new ventures often lose money, the financial position of MoviePass seems extraordinarily shaky. The stock price of its owner, Helios and Matheson Analytics, Inc. (HMNY), had dropped to .0158 per share as of December 21, 2018.

The business case for MoviePass is typical in that uncertainty plays an important role in its financial picture and prospects. MoviePass charged its customers the $9.95 per month for virtually unlimited admissions, while it paid the theatres their full admission prices (averaging $8.97 in 2017) as compensation. Profits depend critically on how frequently its subscribers go to theaters—and even though the company likely had done market research to understand this frequency for the people that would buy the MoviePass, their *actual* behavior and frequency of visiting the movies was essentially uncertain.

Average movie attendance in the overall US population in 2017 was 0.77 movies per person per month (based on total box office revenues of $12 billion and 1.3 billion admissions for the year). Thus, if MoviePass subscribers view movies at the same rate as this average, there would be a "profit" margin of over $3 per subscriber per month

($9.95 − 0.77 * $8.97 = $3.04), which might cover its other operating expenses (e.g., marketing, corporate overhead, etc.). Additional profits were anticipated to be generated by obtaining discounted pricing from theatres and monetizing data collected from its subscriber base. But, the business case depends critically on the admissions behavior of its subscribers, and since MoviePass didn't know yet which population would subscribe, and if and how they might change their movie-going behavior after becoming subscribers, this was subject to uncertainty.

This business situation is typical in several aspects. Myriad factors can influence what the right pricing decision is, and a model can be useful to provide insight, even though, by definition, a model is a simplification of reality and there are lots of uncertainties that affect the results. In order to provide usable insight for the decision problem, the analyst must decide which factors are important to include in a model and which can be ignored. In this case, the decision concerns the MoviePass subscription price, and we will ignore factors that are not directly germane to this decision (pricing decisions can sometimes be viewed as optimization problems, which is a subject we'll consider in Chapter 7). The value of a spreadsheet model[3] typically lies in its capability to perform "what-if" analyses to inform decision makers (Monte Carlo simulation, introduced in the next chapter, can be thought of as what-if analysis "on steroids"). For example, we would like to estimate the impact of changing the subscription price from $10 per month to $20 per month. But that requires estimating how the price increase would affect the number of subscribers, and how frequently those subscribers would attend movies.

How do we begin constructing such a model? The answer is that we start with a visualization of the problem.

1.2 First Steps: Visualization

Since the integrity, and therefore usefulness, of a spreadsheet model depends on having a clear conceptual map of the appropriate relationships in the model, the question is: Where do you begin? Every spreadsheet model at some point begins with a blank spreadsheet, and the novice modeler often begins with a blank stare. The answer, however, is not to begin in the spreadsheet—it is to start on a piece of paper. It is best to create a visualization of the model you want to build. Various authors have given these visualizations different names: influence diagrams (Cleman and Reilly 2010), influence charts (Powell and Baker 2009), mental models or visual models (Ragsdale 2007), or causal loop diagrams (Morecroft 2007). The intent of these tools is identical—to visualize the parts of the model and their relationships. We will refer to any of these as visualization tools, and they are a critical first step in model building. Indeed, if you are careful in constructing your visualization, as it can serve as a blueprint of your model and can almost automate the building of your spreadsheet model.

There is no standardized form to these visualization tools. We will use a variant of the influence chart[3] included in Powell and Baker (2009). The diagram distinguishes between the following elements, each with their own shape:

- Objectives: use hexagons.
- Decisions: use rectangles.
- Input parameters: use triangles. If these are key uncertainties, use dashed lines for the border of the triangle.
- Calculations: use ovals.

Every shape should have one or more connectors to other shapes where these are appropriate. For example, our spreadsheet model for MoviePass will be based on the influence chart shown in Figure 1.1.

The chart is started on the right side with the objective(s). Our simple first model has one objective, profit per subscriber, but your model may contain more objectives if your problem has multiple items of interest. For example, if you are modeling staffing of a call center, you might be interested in analyzing the average time to process calls as well as the total cost of staffing and operating the call center.

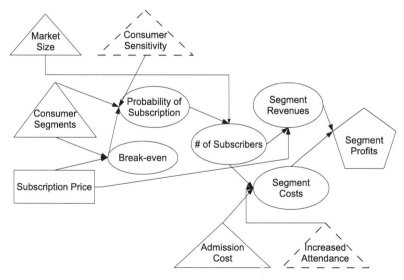

FIGURE 1.1
MoviePass Influence Chart

Building the chart is typically best done by decomposing backward from the objective, one step at a time, until you only have parameters and decisions. Intermediate stages typically involve calculations. To calculate our MoviePass profits, we need to know what the average revenue and cost are for its subscribers. Each of these is calculated based on other things (parameters) in the model. The average cost depends on the average number of theater admissions per subscriber and the average ticket price. Admissions per subscriber depend on customer choice in relation to the admissions distribution: Which consumers will choose to subscribe? We model this based on estimating their probability of subscription, and that is assumed to depend on each consumer comparing their theater attendance behavior with the subscription price.

There are some important things to remember when constructing the visualization:

- Designation of something as a parameter does not mean it is obvious or trivial. It only means that your model will not attempt to explain where this variable comes from. Later models may well replace a parameter with some logic that shows the influences that determine such a parameter. But it is important to start simple, with only the most important factors included in your model.

- Connector lines are very important. In most models, every parameter will directly or indirectly be linked to all objectives, but the connectors represent direct relationships between two elements in the visualization. If there is a connector, then the formula for that cell in the spreadsheet should include reference to the cell it is linked to. If there is no such connector, then there should not be a cell reference.

- Your visualization should not contain any numbers. You are trying to capture essential relationships, to help develop the structure of your model. Numbers belong in the spreadsheet, not in your visualization. A corollary to this is that typically you should keep in mind the type of data that you have available when building your model visualization. In other words, if it is possible to structure your model in a way so that it would rely more on data than on "expert estimates," it is typically better.[4] If there is no data available, it is still be possible to develop some estimate for a parameter, and Monte Carlo simulation (introduced and covered in the next chapter) is often a good way to model such uncertain values.
- Try to be precise when defining terms in your visualization and ensure that they are measurable. For example, the objective "profit per month" is better than simply "profits," as it will more easily translate directly into your spreadsheet. Similarly, a decision of "subscription price" is easier to implement in the spreadsheet than a decision "What should we charge?"

Based on our visualization, our initial spreadsheet model (MoviePass simple1.xlsx) is shown in Figure 1.2.

Based on the average number of theater admissions per month (0.77), it is hard to imagine many people subscribing to MoviePass at a price of $9.95 per month (the breakeven, calculated in cell B16 of the spreadsheet is 1.11 admissions per month), unless people would expect to change their behavior after becoming MoviePass holders. Our model assumes that each consumer segment (corresponding to different levels of initial movie attendance) will base its decision to subscribe to MoviePass on the basis of how its initial attendance compares with the breakeven attendance rate of 1.11 movies per month. The *consumer sensitivity* parameter determines how sensitive consumers are to this breakeven rate. Figure 1.3 shows how the probability of subscription relates to monthly attendance for three levels of consumer sensitivity.

Probability of subscription is shown as an S-shaped curve, generated by the formula $1/(1 + \text{EXP}(-\text{sensitivity} * (\text{average attendance} - \text{breakeven attendance})))$. The graph shows this probability as a function of the number of admissions for three of the behavior parameters. Our functional form causes the probability of subscription to equal 50% when attendance is at the breakeven level, regardless of the sensitivity parameter. As attendance falls below (above) the breakeven level, the probability of subscription falls (rises). The S-shaped curves represent assumed consumer behavior that the probability of subscribing to MoviePass is very low for consumers with very few admissions, but approaches 100% as the number of admissions grows large.

Small sensitivity parameter consumers do not base their subscription behavior on whether their attendance exceeds (or falls short of) the breakeven attendance. In fact, a parameter value of zero causes the subscription curve to be flat, with all consumer segments having a 50% probability of subscription (i.e., subscription does not depend at all on movie attendance).

Large sensitivity parameter values (e.g., = 5) mean that the probability of subscription is highly curved, and the probability of subscription depends heavily on the breakeven attendance rate. This is a key parameter, since it is the subscription of moviegoers who attend relatively infrequently that are most profitable for MoviePass.

Our model provides insight by varying the two consumer behavior parameters, as shown in Figure 1.4.

Conceptual Maps and Models

	A	B	C	D	E	F	G	H	I	J
1	**Parameters**									
2	Attendance distribution		mean (from data)	=SUMPRODUCT(A4:A10,B4:B10)						
3	admissions	fraction								
4	16	0.01	subscription price	9.95						
5	5	0.01	admissions price	8.97						
6	3	0.04								
7	2.1	0.08								
8	0.85	0.19								
9	0.2	0.53								
10	0	0.14								
11										
12	Increase in move going w/MoviePass	0.25								
13	Consumer sensitivity	3								
14										
15	**Calculations**									
16	Break even admissions rate	=D4/D5	admissions/month							
17	Total market size	10000000								
18										
19	Segments:	Freq. w/MoviePass	size	Probability subscription:	# subscriptions:	Revenues	Marg. Costs/subscriber	Marg. Profit/subscriber	Total marg. Costs	Profit/month
20	=A4	=A20*(1+$B$12)	=$B$17*B4	=1/(1+EXP(-$B$13*(B20-$B$16)))	=C20*D20	=E20*$D$4	=A20*$D$5	=$D$4-G20	=G20*E20	=F20-I20
21	=A5	=A21*(1+$B$12)	=$B$17*B5	=1/(1+EXP(-$B$13*(B21-$B$16)))	=C21*D21	=E21*$D$4	=A21*$D$5	=$D$4-G21	=G21*E21	=F21-I21
22	=A6	=A22*(1+$B$12)	=$B$17*B6	=1/(1+EXP(-$B$13*(B22-$B$16)))	=C22*D22	=E22*$D$4	=A22*$D$5	=$D$4-G22	=G22*E22	=F22-I22
23	=A7	=A23*(1+$B$12)	=$B$17*B7	=1/(1+EXP(-$B$13*(B23-$B$16)))	=C23*D23	=E23*$D$4	=A23*$D$5	=$D$4-G23	=G23*E23	=F23-I23
24	=A8	=A24*(1+$B$12)	=$B$17*B8	=1/(1+EXP(-$B$13*(B24-$B$16)))	=C24*D24	=E24*$D$4	=A24*$D$5	=$D$4-G24	=G24*E24	=F24-I24
25	=A9	=A25*(1+$B$12)	=$B$17*B9	=1/(1+EXP(-$B$13*(B25-$B$16)))	=C25*D25	=E25*$D$4	=A25*$D$5	=$D$4-G25	=G25*E25	=F25-I25
26	=A10	=A26*(1+$B$12)	=$B$17*B10	=1/(1+EXP(-$B$13*(B26-$B$16)))	=C26*D26	=E26*$D$4	=A26*$D$5	=$D$4-G26	=G26*E26	=F26-I26
27										
28	Total				=SUM(E20:E26)	=SUM(F20:F26)			=SUM(I20:I26)	=SUM(J20:J26)
29	Market Share				=E28/B17					

FIGURE 1.2
Initial MoviePass Model

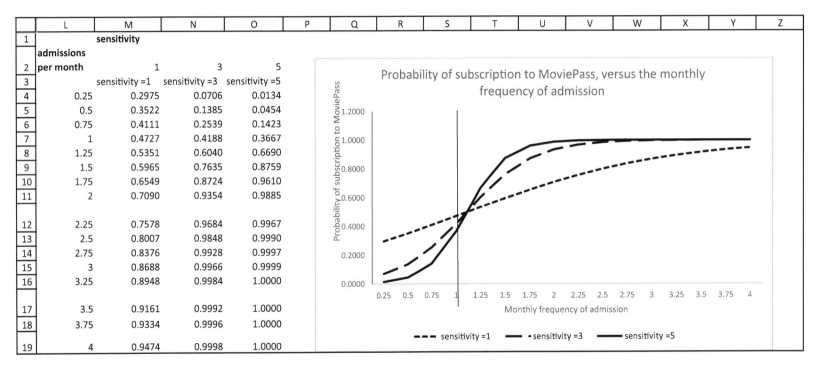

	L	M	N	O
1		sensitivity		
2	admissions per month	1	3	5
3		sensitivity =1	sensitivity =3	sensitivity =5
4	0.25	0.2975	0.0706	0.0134
5	0.5	0.3522	0.1385	0.0454
6	0.75	0.4111	0.2539	0.1423
7	1	0.4727	0.4188	0.3667
8	1.25	0.5351	0.6040	0.6690
9	1.5	0.5965	0.7635	0.8759
10	1.75	0.6549	0.8724	0.9610
11	2	0.7090	0.9354	0.9885
12	2.25	0.7578	0.9684	0.9967
13	2.5	0.8007	0.9848	0.9990
14	2.75	0.8376	0.9928	0.9997
15	3	0.8688	0.9966	0.9999
16	3.25	0.8948	0.9984	1.0000
17	3.5	0.9161	0.9992	1.0000
18	3.75	0.9334	0.9996	1.0000
19	4	0.9474	0.9998	1.0000

FIGURE 1.3
Consumer Sensitivity and the Probability of Subscription

Conceptual Maps and Models

	A	B	C	D	E	F	G	H	I
31									
32			output cell (=J28)			Consumer Sensitivity Parameter		Row input cells C33:I33 (use as B13)	
33	Column input cells (B34:B40) use as B12	($25,075,534)	0.1	0.6	1.1	1.6	2.1	2.6	3.1
34		0%	$8,969,940	-$3,624,857	-$11,455,262	-$17,252,717	-$21,388,601	-$24,243,705	-$26,171,975
35		5%	$8,764,214	-$3,788,326	-$11,575,926	-$17,299,810	-$21,368,939	-$24,172,834	-$26,064,242
36	Increased Moviegoing	10%	$8,566,631	-$3,939,405	-$11,671,170	-$17,313,614	-$21,313,412	-$24,067,028	-$25,924,098
37		15%	$8,377,045	-$4,078,514	-$11,742,876	-$17,298,080	-$21,227,818	-$23,933,253	-$25,758,886
38		20%	$8,195,275	-$4,206,071	-$11,792,878	-$17,256,855	-$21,117,220	-$23,777,283	-$25,574,469
39		25%	$8,021,116	-$4,322,493	-$11,822,950	-$17,193,262	-$20,985,983	-$23,603,872	-$25,375,560
40		30%	$7,854,339	-$4,428,202	-$11,834,785	-$17,110,302	-$20,837,841	-$23,416,925	-$25,165,993

FIGURE 1.4
MoviePass Model with Varying Assumptions

Cells B33:I40 show a data table with the row input cell (consumer sensitivity parameter) as B13 and the column input cell (increased moviegoing) as B12.[5]

The table shows MoviePass profitability as functions of the consumer behavior parameter and the number of theater admissions. It is apparent that the profitability of MoviePass depends critically on the number of times that its subscribers go to the movies—every admission costs MoviePass an average of $9, while its revenue remains fixed at $10 per month. When consumers do not increase their movie going much and are not very sensitive to the breakeven attendance level, the profit opportunity is at its greatest.

At this point we don't have a good way to model either the number of admissions or the consumer behavior parameter using standard Excel tools. All we can do is explore their relevance for profitability. In Chapter 2, we will see that Monte Carlo simulation permits us to better understand how consumer behavior affects MoviePass's profitability.

1.3 Retirement Planning Example

Additional care is required for appropriate visualizations for problems that involve a timed process. It is best to explicitly model the sequence of events or activities and their relationship over time. For example, suppose we want to build a model of a typical retirement planning decision. Consider a 30 year old with no current retirement fund beginning employment at an annual salary of $50,000. Suppose this employee decides to set aside 10% of their pre-tax salary for a retirement fund, an amount to be matched by their employer (up to a maximum of 7%). Assume they anticipate getting annual salary increases that exceed the inflation rate by 2% (we will conduct the analysis in inflation-adjusted dollars, to avoid concern about the value of a dollar in the future). Our employee anticipates retiring at some age between 60 and 65 and wants to see how long their retirement funds will last under these assumptions and for each possible retirement age.

Let us further assume that our employee expects to get a return on their retirement funds that averages 3% above the expected inflation rate, and that after retirement they anticipate needing $50,000 per year (again, in inflation-adjusted or real dollars) to live comfortably. Don't worry about whether these assumptions are correct at this juncture, or even whether they are known for sure. It is the essential structure of the problem we need to capture. Figure 1.5 illustrates a visualization that captures the logic of our problem.

Note that no numbers appear in Figure 1.5. We have not indicated any objectives, nor have we placed the initial fund balance in our diagram. We could do so, beginning with our worker at age 30 and moving through their entire lifetime; however, this would only clutter the diagram. What is important to capture in the visualization is the time structure of the problem, and Figure 1.5 shows the recurring nature of the money going into and coming out of the retirement fund—which depends on our parameter assumptions and whether retirement age has been reached. Figure 1.5 provides a base for constructing a spreadsheet model, which is shown in Figure 1.6 (the spreadsheet spans 70 years, but we have hidden most of the rows in the display—the complete spreadsheet, Retirement1.xlsx, is available on the book website).

These are the salient features represented in this model, listed by the cell address:

A15: Rather than inserting the value 30 for our subject's current age, we have linked it to a parameter representing the current age. This permits the spreadsheet to be easily adapted to a different individual's circumstances. Similarly, the beginning fund balance, cell C15, can be easily changed in the parameter section without needing to change anything in the calculation section of the spreadsheet.

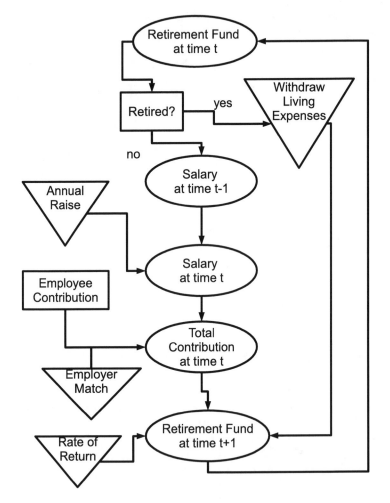

FIGURE 1.5
Retirement Visualization

B15: This formula produces a 0 or 1 value indicating whether the person has retired. It is calculated by comparing the current age with the chosen retirement age (which is a decision).

C15: Initially sets the parameter for the starting fund. Thereafter, equals the ending balance from Column J on the prior row (e.g., C16 = J15).

D15: This calculates the return on the fund balance from the previous year, using the assumed rate of return. We are assuming that returns are earned at the end of each year.

E15: The annual salary is the product of the initial salary and the appropriate number of years of growth at the assumed annual raise. Note also that we multiply this by a factor (1-B15) that results in a salary of 0 when the person is retired or the calculated salary when they are not retired (the same result could be obtained with an appropriate IF function).

F15, G15: These cells calculate the employee and employer contributions to the retirement funds, respectively. The employer's match is limited to equal the employee contribution, or 7%, whichever is lower. H15 simply sums these two contributions. Note that contributions will be zero after retirement since the salary will be zero.

I15: This cell calculates the post-retirement withdrawals from the account. The use of the factor B15 in the formula ensures that no money is withdrawn prior to retirement, since B15 = 0 at that time.

J15: The end-of-year fund is the sum of the initial fund, the annual return, and the employee and employer contributions

	A	B	C	D	E	F	G	H	I	J	K
1	Parameters										
2	Current Age	30									
3	Salary	50000									
4	Annual Raise	0.02									
5	Employee Contribution	0.1									
6	Employer Match	0.07									
7	Rate of Return	0.03									
8	Retirement Living Expenses	50000									
9	Initial Fund Balance	0									
10	Decisions		Objective (year funds run out)								
11	Retirement Age	65	=B2+SUM(K15:K85)								
12											
13	Calculations										
14	Age	Retired?	Fund at Start of Year	Return on Fund	Salary	Employee Contribution	Employer Match	Total Contributions	Withdrawals	Fund at End of Year	fund still available?
15	=B2	=IF(A15>=$B$11,1,0)	=$B$9	=$B$7*C15	=(1-B15)*($B$3*(1+$B$4)^(A15-30))	=$B$5*E15	=IF($B$5>7%,7%,$B$5)*E15	=F15+G15	=B15*$B$8	=C15+D15+H15-I15	=IF(J15>0,1,0)
16	=A15+1	=IF(A16>=$B$11,1,0)	=J15	=$B$7*C16	=(1-B16)*($B$3*(1+$B$4)^(A16-30))	=$B$5*E16	=IF($B$5>7%,7%,$B$5)*E16	=F16+G16	=B16*$B$8	=C16+D16+H16-I16	=IF(J16>0,1,0)
17	=A16+1	=IF(A17>=$B$11,1,0)	=J16	=$B$7*C17	=(1-B17)*($B$3*(1+$B$4)^(A17-30))	=$B$5*E17	=IF($B$5>7%,7%,$B$5)*E17	=F17+G17	=B17*$B$8	=C17+D17+H17-I17	=IF(J17>0,1,0)
83	=A82+1	=IF(A83>=$B$11,1,0)	=J82	=$B$7*C83	=(1-B83)*($B$3*(1+$B$4)^(A83-30))	=$B$5*E83	=IF($B$5>7%,7%,$B$5)*E83	=F83+G83	=B83*$B$8	=C83+D83+H83-I83	=IF(J83>0,1,0)
84	=A83+1	=IF(A84>=$B$11,1,0)	=J83	=$B$7*C84	=(1-B84)*($B$3*(1+$B$4)^(A84-30))	=$B$5*E84	=IF($B$5>7%,7%,$B$5)*E84	=F84+G84	=B84*$B$8	=C84+D84+H84-I84	=IF(J84>0,1,0)
85	=A84+1	=IF(A85>=$B$11,1,0)	=J84	=$B$7*C85	=(1-B85)*($B$3*(1+$B$4)^(A85-30))	=$B$5*E85	=IF($B$5>7%,7%,$B$5)*E85	=F85+G85	=B85*$B$8	=C85+D85+H85-I85	=IF(J85>0,1,0)

FIGURE 1.6
Retirement Spreadsheet Model

	A	B	C	D	E	F	G	H	I	J	K
1	**Parameters**										
2	Current Age	30									
3	Salary	$50,000									
4	Annual Raise	2%									
5	Employee Contribution	10%									
6	Employer Match	7%									
7	Rate of Return	3%									
8	Retirement Living Expenses	$50,000									
9	Initial Fund Balance	$0									
10	**Decisions**		**Objective** (year funds run out)								
11	Retirement Age	65	83								
12											
13	**Calculations**										
14	Age	Retired?	Fund at Start of Year	Return on Fund	Salary	Employee Contribution	Employer Match	Total Contributions	Withdrawals	Fund at End of Year	fund still available?
15	30	0	0	0	50,000	5,000	3,500	8,500	0	8,500	1
16	31	0	8,500	255	51,000	5,100	3,570	8,670	0	17,425	1
65	80	1	147,976	4,439	0	0	0	0	50,000	102,415	1
66	81	1	102,415	3,072	0	0	0	0	50,000	55,488	1
67	82	1	55,488	1,665	0	0	0	0	50,000	7,152	1
68	83	1	7,152	215	0	0	0	0	50,000	-42,633	0
69	84	1	-42,633	-1,279	0	0	0	0	50,000	-93,912	0
70	85	1	-93,912	-2,817	0	0	0	0	50,000	-146,729	0
71	86	1	-146,729	-4,402	0	0	0	0	50,000	-201,131	0
72	87	1	-201,131	-6,034	0	0	0	0	50,000	-257,165	0
73	88	1	-257,165	-7,715	0	0	0	0	50,000	-314,880	0
74	89	1	-314,880	-9,446	0	0	0	0	50,000	-374,326	0
75	90	1	-374,326	-11,230	0	0	0	0	50,000	-435,556	0
76	91	1	-435,556	-13,067	0	0	0	0	50,000	-498,623	0
77	92	1	-498,623	-14,959	0	0	0	0	50,000	-563,582	0
78	93	1	-563,582	-16,907	0	0	0	0	50,000	-630,489	0
79	94	1	-630,489	-18,915	0	0	0	0	50,000	-699,404	0
80	95	1	-699,404	-20,982	0	0	0	0	50,000	-770,386	0
81	96	1	-770,386	-23,112	0	0	0	0	50,000	-843,497	0
82	97	1	-843,497	-25,305	0	0	0	0	50,000	-918,802	0
83	98	1	-918,802	-27,564	0	0	0	0	50,000	-996,366	0
84	99	1	-996,366	-29,891	0	0	0	0	50,000	-1,076,257	0
85	100	1	-1,076,257	-32,288	0	0	0	0	50,000	-1,158,545	0

FIGURE 1.7
Retirement Model Base Case

during the year, minus the withdrawals (becomes the starting balance in the next year, C16).

K15: This cell indicates whether the end-of-year fund is positive. It is used to calculate the year that the retirement fund is exhausted in cell C11. The formula in C11 adds the initial age to the number of years that show a positive fund balance, thus yielding the age at which the funds are gone (our objective).

Note that all numbers appear in the parameter section and only formulas appear in the calculation section. This is an important good practice in spreadsheet modeling and permits the spreadsheet model to be easily adapted to varying circumstances without needing to reconstruct the model.[6] Note also our use of relative and absolute cell addressing. Use of $ in front of a row number or column letter will freeze that number or letter—thus, it will not change when copied (called *absolute* addressing). Absence of the $ means that the row number will be changed when copied downward or the column letter will be changed when copied rightward in the spreadsheet (called *relative* addressing). Appropriate use of absolute and relative addressing greatly simplifies the construction of models by permitting many formulas to be entered once and copied throughout the spreadsheet. The flip side is that improper application of relative and absolute addressing is also one of the most common errors made in building spreadsheet models.

The value of this model lies in its ability to be used for a variety of what-if analyses. The base case for our model is shown in Figure 1.7 which produces an age of 83 for when the retirement fund is exhausted if the person retires at age 65.

Our model can be used to examine the sensitivity of the exhaust year to our two decisions: the employee annual contribution to the fund and the chosen retirement age. We do this by using the Data Table command in Excel. The result is the table shown in Figure 1.8.

As Figure 1.8 shows, the retirement funds are expected to run out as early as age 66 (for the earliest retirement age and the lowest employee contribution factor) or as late as age 91 (if the worker works until age 65 and contributes 15% of their annual salary). If our worker wishes to have the fund last until their expected lifetime (say, 83 years

	A	B	C	D	E	F	G	H	I	J	K	L	M
1							Empoyee Contribution Factor						
2		83	5.00%	6.00%	7.00%	8.00%	9.00%	10.00%	11.00%	12.00%	13.00%	14.00%	15.00%
3	Retirement Age	60	66	68	70	70	71	72	73	74	75	76	77
4		61	68	70	71	72	73	74	75	76	77	78	80
5		62	69	71	73	74	75	76	77	78	80	81	82
6		63	71	73	75	76	77	78	79	81	82	83	85
7		64	72	75	77	78	79	80	82	83	85	86	88
8		65	74	76	79	80	81	83	84	86	87	89	91

FIGURE 1.8
Retirement Fund Exhaust Year Sensitivity

old, for a female in the United States), then this result can be obtained by working until age 64 with a contribution rate of 12% (or working until age 63 with a contribution rate of 14%).

Our model is quite flexible and can produce several insights, but it is severely limited by the fact that there are a few uncertainties not represented in the model. This is a good application for Monte Carlo simulation, and thus we will begin the next chapter with further refinements to this retirement model.

1.4 Good Practices with Spreadsheet Model Construction

If you are careful in constructing the visualization, building the spreadsheet model will be much easier and efficient. Organization is provided by the shapes of the boxes: put parameters in one section of the spreadsheet (possibly divided into those that are certain and those that are uncertain), decisions are in a separate section, calculations are in a separate section, and results and/or objectives are in the final output section (frequently also put near the top for ease of viewing). If the spreadsheet is large, then these sections may better be placed on separate worksheets, and different portions of the calculations may require separate worksheets. As an overall recommendation, when building a spreadsheet model, you should always try to start small. It is easy to expand or enhance a model once a simple version is working.

Typically, spreadsheet models would contain only numbers and labels in the parameter and decision sections (or inputs for probability distributions for uncertain parameters, as discussed in the next chapter). The calculations and objectives would consist purely of formulas—with no numbers whatsoever, only cell references. Any time you want to use a different set of input parameters, you simply replace its value in the parameter/input section, and it will then be used (because of the linking) in the calculation part of the model. A main advantage of this is that you can easily perform what-if analyses by modifying the parameter values without worrying about whether you have appropriately modified the formulas. It also allows you (and other users of the model) to much more quickly and easily review your models for accuracy, logic, and input assumptions.

Beginning with a careful visualization and following these rules will greatly help the task of building a sound spreadsheet model. We don't mean to suggest it is always easy—there is often still a significant task in finding the right formulas to use in the calculations and identifying the correct structure for your visualization. The latter is what makes spreadsheet modeling when you're just getting started sometimes challenging. Models, as with maps, simplify reality by omitting details. Their usefulness depends on the uses to which they will be put and the available information. One of the best ways to learn how to do this is through repeated application and trial and error. Visualizations can often cure the problem of the blank stare, or "Where do I begin?" In addition, there are some ground rules to help you prevent a lot of common errors.

1.5 Errors in Spreadsheet Modeling

Spreadsheets have become a ubiquitous business tool. The first spreadsheet program, VisiCalc, was released in 1979. Lotus 123, released in 1983, became the first dominant spreadsheet program, but Excel quickly became the standard after its release. By 1989, it was the dominant spreadsheet program. As its use has grown, so has the realization that many (indeed, most) spreadsheets in use contain errors, and many of these are serious enough to have resulted in decisions that differ from what would have been done if the spreadsheet had been correct.

Considerable research has been conducted into the prevalence, detection, and prevention of these spreadsheet errors. Estimates range from more than 50% of all spreadsheets having errors of some type to virtually all spreadsheets having errors. One study even found that errors are not reduced with experience in building spreadsheets. Three of the main research centers (with extensive websites and links to research) are:

- European Spreadsheet Risks Interest Group (www.eusprig.org)—in particular, they maintain a list of "horror stories" documenting the extensive costs associated with spreadsheet errors.
- Spreadsheet Research (panko.shidler.hawaii.edu/SSR/index.htm).
- Spreadsheet Engineering Research Project (mba.tuck.dartmouth.edu/spreadsheet/).

These sites are worth visiting. They have links to research papers as well as best practices and tools to assist with identifying and preventing mistakes in spreadsheets. Several tools are commercially available, and Excel has some built-in tools for identifying spreadsheet errors. Trace Dependents and Trace Precedents (found on the Formula toolbar) can be used to identify which cells in a spreadsheet are linked to other cells. This can be particularly useful if used in conjunction with a model visualization, such as an influence chart.

For example, Figure 1.9 shows a section of the MoviePass spreadsheet with Trace Precedents set on the probability of subscription cell for the highest use segment (D20). It shows that three cells are used in the formula (consumer sensitivity, breakeven, and frequency w/MoviePass), and the influence chart in Figure 1.1 shows that

	A	B	C	D
1	Parameters			
2	Attendance distribution		mean (from data)	0.7655
3	admissions	fraction		
4	16	0.01	subscription price	$9.95
5	5	0.01	admissions price	$8.97
6	3	0.04		
7	2.1	0.08		
8	0.85	0.19		
9	0.2	0.53		
10	0	0.14		
11				
12	Increase in move going w/MoviePass	25%		
13	Consumer sensitivity	3		
14				
15	Calculations			
16	Break even	1.11	admissions/month	
17	Total market size	10000000		
18				
19	Segments:	Freq. w/MoviePass	size	Probability subscription:
20	16	20.0	100000	1.000

FIGURE 1.9
Trace Precedents

there are three arrows pointing to the calculation of probability of subscription. The influence chart and the spreadsheet should match: links in the influence chart should correspond to the presence of a cell in a formula.

The Data Validation (found on the Data toolbar) tools permit you to restrict the types of values (e.g., numbers, dates, etc.) that can be used in particular cells as well as the range of acceptable values. Data Validation will circle cells with errors so that they can be further investigated. Automated Error Checking will identify potential cells with errors, such as when the cell contains a formula that does not look like the formulas in adjacent cells. This can help locate the (all too common) copy and paste errors.

A number of commercial products are available that go beyond these simple tools. One of the best practices for building spreadsheets is to use such tools, but they can only identify certain types of errors.

A survey of the types of spreadsheet errors can be found in Panko and Aurigemma (2010).[7] They classify errors first into two types: culpable and blameless. There are methods for identifying and preventing culpable errors, but these are similar to methods used to reduce other types of intentional errors, such as employee stealing, sabotage, and so on. These are primarily organizational issues rather than technical, although some may be detected with the same tools that are used to detect blameless errors.

Blameless errors are divided into quantitative and qualitative errors, with quantitative errors further subdivided into planning errors and execution errors. These can be further delineated into a number of categories, such as copy and paste errors, overconfidence, archiving errors, and so on. Readers should spend some time at the above sites to get a good sense of the frequency, types, and consequences associated with the myriad kinds of spreadsheet errors. One observation that is not generally emphasized is that there is no hierarchy to these errors. While we suspect that qualitative errors are in some ways "more serious" than quantitative errors, the consequences cannot be easily ranked. A single copy and paste error can have more practical significance than an important conceptual error, particularly in a complex spreadsheet. It is probably easier to detect the quantitative errors since they are more amenable to automated auditing tools, but this does not mean that the errors are less serious.

1.6 Decision Analysis

Decision analysis (DA) is the discipline of formally addressing important decisions in business, the public sector, or even in personal decision making. DA is a wide and interdisciplinary field and comprises several fields, including psychology, economics, and management science. Especially when the outcomes of decisions are uncertain, a structured and thorough approach to making the decision can be very helpful and can help increase the quality of a decision.

While this book focuses on an important aspect of DA (the quantification of decision-alternatives, and their respective risks), we would also encourage readers to also learn about other important aspects of what it takes to make good decisions. There are lot of good books and (online) resources to learn about DA. Two good starting points would be the Society of Decision Professionals (https://www.decisionprofessionals.com/) and the INFORMS association (https://www.informs.org/).

1.7 Conclusion: Best Practices

At the outset, we emphasize that there is no "one size fits all" methodology to follow in spreadsheet modeling. Thus, we humbly submit the following "recommendations" rather than rules:

- Begin with a visualization of the problem to be modeled, not in a blank spreadsheet. More experienced users will often skip this step, but when faced with novel and complex problems will usually start here. If you can't picture the logical structure of your spreadsheet, then it is not likely to emerge from a spreadsheet.[8]

- Do not begin by entering data in a blank spreadsheet. Most problems have some input data (e.g., a set of prices, a list of characteristics, etc.) that can readily be placed in a spreadsheet. Novice modelers struggle for a place to begin, so are tempted to enter this data, knowing that they will use it later on. However, the way the data is entered should be governed by the logical structure of the problem, not by the order in which the numbers are encountered. Once the data is entered in the spreadsheet, users are then forced to work around the (arbitrary) order in which it was entered. It is always better to have a logical structure in mind, and then place the data in a form and format in which it belongs.

- Make simplifying assumptions: Try to pare the problem down to its essential core. Spreadsheets are natural environments for starting simple and relatively easily expanding your model (since copying cells is so easy), so it is typically best to start small and only expand after you have captured the essential relationships in your model and learned which parts of the model are most important.[9]

- It may help to first look at available data to help guide the construction of your model, but do not worry too much about missing information. If a factor is important to your model, it should be in your model. It is always better to do additional research to find estimates for missing data that is critical than to leave it out of your model because it wasn't available to you initially. Alternatively, consider constructing the model differently so that unknown parameters may be avoided.

- Separate your model into sections: Parameters, decisions, calculations, and outputs. Strive to have numbers (or distribution parameters for simulation) in the parameter and decision sections, and only formulas in the calculations and output/objectives sections.

- Document extensively: Insert comments or provide columns for comments that list the source for data and formulas as well as dates for when these are obtained. The more a spreadsheet is to be used by other people, the more important this becomes since subsequent users of the spreadsheet are not likely to be familiar with the source material. Spreadsheets will outlast their originators, so this documentation will be lost if it only exists in the mind of the original creator of the spreadsheet.

- Quality control is key: Test each formula as you enter it. Try extreme values to make sure it works as intended. Use appropriate auditing tools. Be especially careful when copying and pasting data that you use the correct target range of cells (an easy thing to check is whether the last cell you paste into has the correct formula in it). Pay attention to relative and absolute cell addressing. Organizations should strive to establish quality control procedures for spreadsheet development. If you work in an organization, do you have established procedures? Most do not, but should. Revisit the "horror stories" on the EUSPRIG website if you need to be convinced.

- Try to keep your formulas simple: If you find yourself entering a lengthy formula, try breaking it into several steps (better to have five cells, each with small formulas, than one cell with a huge mega-formula). Novice modelers impress themselves (or others) when they build a complex formula and it does what they want. While this may appear impressive, it is a typically a poor modeling habit. Spreadsheets have plenty of room and it is easier to troubleshoot your formulas if they are small than if they are large.
- If you find yourself with calculation columns without headers or values entered without labels, go back and find appropriate narrative labels for them. Commonly, unlabeled columns reflect that the modeler wasn't really sure what the formula was intended to accomplish. It will also be easier for other users of your spreadsheet if everything has a descriptive title or label to inform them of what its purpose is.
- Build graphs in your spreadsheet to illustrate the effects of different input data on your objective cells. This is an excellent way to communicate your results as well as a good way to make sure your model is working as intended. The graph should make intuitive sense. If it looks odd, then it may mean something is wrong with your model. For example, if your results show no sensitivity to a key parameter (or one that you thought was important), then quite possibly you have failed to use that parameter in the correct manner.
- If you work on building models with others (and we recommend that you do), make sure you establish effective collaboration methods. It is not unusual to find two people making the same error in a spreadsheet if they work together. A good technique would be to jointly work on a visualization of a problem and then to separately create the spreadsheet prototype from your visualization. Comparison of models and results is usually informative.
- Protect against overconfidence! Spreadsheet modelers are prone to believe in the results of their models, and to do so more strongly the more effort they expend in their creations. Research confirms these tendencies toward unwarranted overconfidence. Modelers also tend to be overconfident in the values they place in their spreadsheets. Often a initial input value is chosen as a temporary placeholder, based on limited information, but it is subsequently perceived as more certain than it actually is. Possible protection against this is to use Monte Carlo simulation to represent the uncertainty about key parameter values. This is the subject for the next chapter.

Exercises

For each of the following exercises, develop an appropriate visualization and then build a working spreadsheet to examine a base case. The base case should simplify the problem to its essential components and should not consider any uncertainty (you will have plenty of opportunities to do this in the next chapters).

Exercise 1.1 Health Insurance Choices

Health insurance decisions are important and complex. Many plans are available, and they differ in numerous dimensions—virtually all are expensive. Your task is to build a model that will help individuals choose an appropriate plan from among a number of options. For this exercise, Table 1.1 provides data extracted for family coverage.[10]

Assume that this family's usage patterns are representative of the average in the population, as shown in Table 1.2.[11]

TABLE 1.1

Health Care Plan Costs and Features

Plan	Monthly Cost	Family Deductible	Out-of-Pocket Maximum	Co-Pay for Inpatient Benefits	Office Visit Co-Pay (after Deductible)
Bronze	$1616	$9450	$13,300	40%	40%
Silver	$1922	$6600	$15,800	50%	$10 (primary), 50% specialist
Gold	$2050	$2250	$15,800	30%	$15 (primary), $50 specialist

TABLE 1.2

Family Health Care Usage Patterns

Family Member	Average Office Visits	Average Inpatient Costs
45-year-old female	0.88 primary care 3.29 specialist	$1868
45-year-old male	0.66 primary care 2.46 specialist	$1868
15-year-old male	1.23 primary care 0.91 specialist	$772
12-year-old female	1.23 primary care 0.79 specialist	$772

Assume that primary care outpatient office visits cost an average of $150 and outpatient specialist visits average $250. Build a model to determine which plan is cheapest for this family.

Exercise 1.2 Electric Vehicles

Consumers are becoming more sensitive to concerns about global warming, as well as concerns about their spending on gasoline. In 2016, all electric vehicles accounted for around 1% of US auto sales, but these are projected to reach 22% of sales by 2025. The initial cost for electric vehicles is generally higher than for gasoline-powered cars. Build a model to estimate the breakeven additional cost for a hybrid vehicle under the following assumptions: the automobile is driven 13,476 miles per year, lasts for 12 years, and two types of engines are available:

- Conventional gasoline engine: averages 37.7 miles per gallon
- All-electric engine: averages 4 mi/kwh on electricity

Assume that the car will be driven the same number of miles, regardless of engine type. Examine the breakeven extra cost (in terms of net present value) for the electric engine for gasoline prices of $2.46/gallon and electricity prices of $0.13/kwh (approximate US averages, end of year 2018). In other words, how much more would a consumer be willing to pay for an electric car, based solely on fuel cost savings? Assume the discount rate is 5%.

Exercise 1.3 Investing in Education

Choosing to go to college is one of the biggest lifetime decisions people make. Economists view this, at least partially, as investing in human capital. Ignoring the consumption value of college, there is certainly an expectation that income will be higher as a result of obtaining a degree. The data bears this out: mean annual earnings for 25–29-year olds with a high school diploma are $35,256—with a college degree, these increase to an average of $59,124. Professional degree holders average $89,960.[12] Income rises faster with education as well: in the 55–59-year-old age group (the peak earning age), high school graduates average $49,597 of income, college graduates average $89,283, and professional degree holders average $168,825. Higher income levels will be subject to higher taxes. Use 4%, 8%, and 13% as the average tax rates that apply to high school graduates, college graduates, and professional degree holders, respectively.

Additional data concerns the cost of college. Net tuition and fees (after financial aid is taken into account) average $3740 (per year) at public colleges and $14,600 at private colleges (65% of college students are enrolled at public colleges, 35% at private ones). Assume that college takes 4 years. Professional education averages $30,000 per year and requires an additional 2–4 years (use 3 years as an average in this exercise). For textbook expenses, use an average of $750/year. We ignore the room and board costs as food and lodging are necessary regardless of whether one attends college. But do include the opportunity cost (lost earnings) of attending college rather than working full-time.

A financial measure of investing in college (or professional schooling) is the net present value (NPV). Estimate the NPV of investing in a college degree (taking 4 years), assuming a discount rate of 7%. Also, estimate the NPV of investing in professional school (say, an MBA), taking an additional 3 years. Assume a person aged 20 years and a retirement age of 65 years.

Exercise 1.4 Customer Lifetime Value

Customer lifetime value (LTV) is a concept developed to consider the fact that it generally costs something to acquire customers and that the profits rely on retaining them over time. Consider the hypothetical data in Table 1.3 concerning customer profitability over a 3-year time frame:

Assume that your company has 100,000 initial customers. Estimate the LTV, which is defined as the cumulative net present value of the profits divided by the original number of customers. Use a discount rate of 15%.

Exercise 1.5 Network Economics

Communications technologies usually exhibit network effects: their value increases more than proportionally with the number of users. For example, instant messaging services are more valuable the more people you can exchange messages with. Under such conditions, market growth can be self sustaining, provided that enough users adopt the service. But such a market must first exceed critical mass: the number of users required in order for the user base to grow. Failure to achieve critical mass may result in the user base unraveling. Figure 1.10 illustrates the situation.

TABLE 1.3

Customer Lifetime Value

Year	Retention Rate	Orders/Year	Average Order	Cost of Goods Sold	Acquisition/ Marketing Costs
1	60%	1.8	$85	70%	$60
2	70%	2.5	$95	65%	$20
3	80%	3.2	$105	65%	$20

The curve in Figure 1.10 represents the value to the marginal user under the following conditions. Each user has an intrinsic value that represents the value to them, assuming that everybody else is also a user. Since people differ, these intrinsic values will differ across people. In this exercise, assume that the intrinsic values are given by $100(1 - F)$, where F is the fraction of the population that uses this service. This means intrinsic values start at $100 (for the person with the highest valuation) and range down to $0 for the person with the lowest value.

The actual realized value for each user is given by their intrinsic value multiplied by F: if everybody subscribes ($F = 1$), then each user gets their intrinsic value. If nobody subscribes, then nobody gets any value. If 50% subscribe, then everybody gets half of their intrinsic value. So, the actual realized value is given by $100F(1 - F)$, and this results in a curve such as what is shown in Figure 1.10.

Assume the marginal cost is constant with the number of users and equals $5. Figure 1.10 has three equilibrium points. To the left of the critical mass point, subscribers will have values less than the cost and will stop subscribing: eventually driving the equilibrium to zero. Critical mass represents an unstable equilibrium. As soon as the number of users rises above critical mass, values will exceed costs and the subscriber base will grow until the saturation level is reached. If the user base ever exceeds the saturation level, then there will be users with values less than the cost and they will eventually drop their subscriptions.

Determine the critical mass and saturation levels under these conditions.

Exercise 1.6 Peak Load Pricing

Electricity demand varies over time. It is expensive to maintain generating capacity sufficient for meeting the peak demand—especially if that capacity sits idle the rest of the time. In this exercise, assume that there are two hourly demand levels, peak and off-peak. Both are influenced by price. For the peak demand, assume $Q = 125 - P$, while off-peak demand is given by $Q = 70 - 1.2P$. Peak demand is experienced 8 hours each day (for example, from 10 AM until 6 PM). Assume that capacity costs, expressed on a daily basis, equal 600 for each unit of capacity (one unit of capacity can serve one unit of demand per hour—but note that capacity can be reused across hours, that is, it is not used up in production). There are no other costs to consider.

Compare the profitability of two pricing schemes: first, a single price of 50 for every hour of the day, and second, a peak load pricing scheme that charges 75 during peak hours and 25 during off-peak

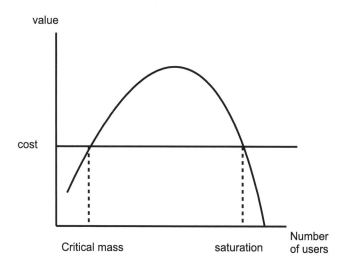

FIGURE 1.10
Network Effects

hours. Also, show the demand profile over a typical day under each of these pricing structures.

Exercise 1.7 Project Management

A new product development cycle involves several interrelated activities. Some activities can be pursued simultaneously, while some can only be undertaken after others are completed. Table 1.4 shows these stages for a typical new product.

What is the expected total time to completion for this project?

Exercise 1.8 Hollywood Finance

The worldwide movie industry is worth over $100 billion annually and the focus of a good portion of popular culture. Develop a model to estimate the profitability for an individual film, using the following information[13]:

- An average US movie costs $72 million to produce and $37 million to promote. Assume this takes place over a 6-month time period.
- Domestic box office receipts average $19.6 million and are earned during the first month of a film's release.
- Foreign box office receipts average about the same as domestic box office receipts, earned during the first 2 months of release.
- Home video rentals average 80% of the total domestic box office receipts, and occur an average of 1 month after domestic theatre runs are finished.
- Home video sales average 220% of the domestic box office receipts and occur at the same time as the rental income.
- Premium pay TV revenues average $7 million/film, occurring over the 12-month period after video sales end.
- Network and free cable television revenues average $7 million/film, earned over the 12-month period following pay TV.
- Syndication averages around $2 million total, earned over the final 5-year period.

Develop a visualization and a spreadsheet model to estimate the average profitability of a US movie. Measure profitability by the internal rate of return.[14]

TABLE 1.4

Project Activities List

Activity	Description	Required Predecessors	Duration
A	Product Design	None	5 months
B	Market Research	None	6 months
C	Production Analysis	A	2 months
D	Product Model	A,C	3 months
E	Sales Brochure	A	2 months
F	Cost Analysis	C,D	3 months
G	Product Testing	D	4 months
H	Sales Training	B,E	2 months
I	Pricing	F,H	1 month
J	Project Report	F,G,I	1 month

Notes

1. This case is purely illustrative, based on publicly available data. It should not be interpreted as a definitive business case, and it is not presented as an accurate portrayal of MoviePass finances and opportunities.
2. The evolution of MoviePass and its pricing was widely reported in the media. For example, see "MoviePass, Rattled by a Rough Year, Tries to Change Its Ways." *New York Times*, December 5, 2018.
3. Models can be developed in many ways, ranging from "back of an envelope calculation" to models in software packages like R and Python. In this book, we'll call them spreadsheet models, but that's only because the example models with the book all come in the form of a spreadsheet, and all can also be put into many other packages with the same results and insights.
4. To add to this, sometimes it is preferable to design a model that does not depend on parameters that are unknown, but instead design an alternative model (possibly a simpler model, or a model that looks at the problem from a different perspective) where parameters with unknown values can be avoided.
5. The Data Table in Excel is an array formula constructed as follows: put values for the consumer sensitivity parameter in cells C33:I33 and values for increased movie going in cells B34:B40, and put the formula you want to focus on, total profitability (J28), in cell B33, as = J28. Then highlight the whole table B33:I40, and select Data Table, found under the Excel Data menu, What If analysis, Data Table. Enter B13 as the row input cell and B12 as the column input cell and Enter. The table will show up with {} as an array formula. Figure 1.3 shows the result.
6. In fact, in many of the models we build, we often give all inputs special formatting so that it is very clear to the user which cells can be updated with new data or assumptions and which cells are calculations or results.
7. Aurigemma, S. and Panko, R. R. 2010. "The Detection of Human Spreadsheet Errors by Humans versus Inspection (Auditing) Software." arXiv:1009.2785.
8. Also, visualization can help explain the model and its results, *after* it has been developed.
9. Later in the book, we'll discuss tornado charts, which can be particularly useful to understanding this.
10. Data comes from HealthCare.gov for three "comparable" plans in the Chicago area (zip code 60601) for 2018. All plans are preferred provider plans (not HMOs) priced for a family of four with two 45-year-old parents, a 15-year-old boy, a 12-year-old girl, and family annual income of $150,000. Table 1.1 compares the major features of these plans—dozens of details (potentially important ones) have been omitted, for example, prescription drug discounts, eye exams, special provision for x-rays, etc.
11. Data for outpatient visits is adapted from the National Ambulatory Medical Care Survey: 2015 State and National Summary Tables, www.cdc.gov/nchs/fastats/physician-visits.htm. Inpatient cost data is adapted from Health Expenditures by Age and Gender from the Centers for Medicare & Medicaid Services.
12. Income data comes from the US Bureau of Labor Statistics, 2018. Tuition rates come from College Board, Annual Survey of Colleges. Unemployment rates (used in Exercise 2.3) come from the Bureau of Labor Statistics.
13. These are stylized facts, based to some extent on data in Young, S. M., Gong, J. J. and Van der Stede, W. A. 2010. "The Business of Making Money with Movies." *Strategic Finance*, 91(8).
14. The internal rate of return is the discount rate that makes the NPV equal to zero. Use the Excel IRR function for this (although, note the caveats about this function in Exercise 2.8). There are technical issues with IRR calculations, problems with the way that Excel calculates the IRR, and theoretical concerns with its use as well (see https://www.mckinsey.com/business-functions/strategy-and-corporate-finance/our-insights/internal-rate-of-return-a-cautionary-tale). Despite these limitations, it remains a widely used tool in finance.

2

Basic Monte Carlo Simulation in Spreadsheets

Learning Objectives

- Understand the importance of uncertainty for model building and in using models to support decision making.
- Learn about Monte Carlo simulation and how it is a good way to consider risk and uncertainty in your models.
- Review the steps of how to build Monte Carlo simulation in Excel spreadsheets.
- Acquire the basic skills of setting up a (Monte Carlo) simulation model and interpreting its results.

2.1 Introductory Case: Retirement Planning

Decisions around retirement planning can be greatly facilitated and supported by using models. People need to make complex and interdependent decisions regarding how much to save for retirement, when to retire, and what lifestyle and spending level to enjoy after retirement. We developed a fairly simple model in Chapter 1, for a 30-year-old person beginning a new retirement account (though our model can readily be adapted to a number of different characteristics). The model in Chapter 1, however, is only considering an "expected case" and of limited use because it omits the reality that there are many uncertainties that are critical to making retirement plans. As the required disclaimer states, "past performance is no guarantee of future results." We all know that future investment returns are uncertain. Similarly, we know that future health care expenses and our employment status are also uncertain and this should be accounted for in our planning.

We will use the retirement model from Chapter 1 to show how it can be enhanced to account for these uncertainties and how the results can increase our understanding of retirement decisions.

2.2 Risk and Uncertainty

There are many definitions of risk and they are often ambiguous. In his classic work, Knight (1921) distinguished between risk and uncertainty wherein the former is something measurable and the latter is not.[1] More recently, Taleb (2007) has popularized the notion of important risky events for which the probability of occurrence is difficult, if not impossible, to measure.[2] He argues that these are the truly important risks, but they are ones for which there is little or no evidence on which to base an analysis. For our purposes, given the diversity of opinions on the right definition, we will not distinguish among the various definitions and we will not confine ourselves to measurable risks.

The presence of uncertainty means that there are probabilities attached to different potential outcomes. We will sidestep for now the issue of how such probabilities and uncertainties may be estimated from data or estimated by experts (Chapter 3 addresses this), but just posit that probabilities can be attached to the various outcomes. A *probability distribution* can be described as the set of possible outcomes, each with its associated probability of occurring. There are two types of probability distributions, discrete distributions for outcomes that can only take on a discrete set of values, and continuous for outcomes that can be any value within a certain range (for some probability distributions, this "range" may even go from minus infinity to plus infinity). Figure 2.1 illustrates a discrete (the individual bars) and a continuous probability distribution (the smooth curve).

With discrete distributions, all probabilities must range between 0 and 1, and the probability of all potential values must sum to 1. For

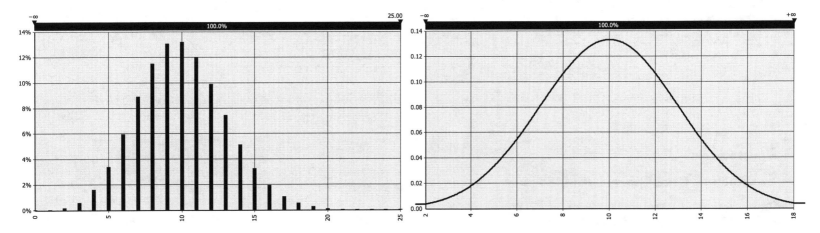

FIGURE 2.1
Probability Distributions

continuous distributions, the area under the curve always must be equal to 1.

We all know that the amount of data that is being produced is increasing very fast. One recent estimate of the amount of data produced daily is 2.5 quintillion bytes of data, with over 90% of the world's data generated in the last 2 years.[3] However, we also know that, even though so much data is now available, when making a forecast, (without allowing for uncertainty) it will likely be wrong! A company forecasting its earnings, a contractor estimating the costs of building a house, or you trying to predict how long it will take to drive from home to work all involve "forecasts" that are uncertain. Whatever the past or present looks like, there is no guarantee that the future will behave the same—in fact, this is why decision makers have a marketable skill. It is precisely the fact that the future cannot be predicted with certainty that makes decisions difficult. No amount of data can change this fact. In the immortal words of Yogi Berra: "The future ain't what it used to be."[4]

When developing models to support decision makers, it would therefore be very useful to include our uncertainty *quantitatively* within the analysis. This will help in getting better insight into what can actually happen and not just what we think is "most likely" going to happen. The way we can do this within a (spreadsheet) model is that we will include uncertainty in the relevant inputs to our model. Such inputs may be based on historical data such as past stock returns or on expert judgments about the uncertainty about future stock returns, or (frequently) on combinations of both. To include such uncertainty about the future, we will need to replace some of the fixed parameters (e.g., our assumption about a rate of return of 7% per year in Chapter 1) with representations (i.e., probability distributions as shown in Section 2.4 below) that take into account that a range of values is possible, and we thus have uncertainty.

2.3 Scenario Manager

We have already been introduced in Chapter 1 to the concept that we can use "what-if" analyses to examine the impact of different parameter values on our objectives. Excel contains a potentially useful tool to do such what-if analysis in a semi-automatic way, called the "Scenario Manager," which is found on the Data ribbon under "What If" Analysis. To see how this tool works, let's return to our MoviePass model from Chapter 1. Figure 1.1 showed that there were two principal uncertain parameters in our model—(1) consumer sensitivity and (2) the increased movie attendance for subscribers. A common approach for modeling key uncertainties is to model pessimistic or optimistic scenarios (sometimes referred to as "worst"- and "best"-case scenarios). For example, we might believe that the pessimistic scenario would involve consumer sensitivity of 5 (rather than the base case of 3) and relatively large increase in movie attendance for subscribers of 50% (rather than the base case of 25%). We can define this pessimistic scenario with the Scenario Manager. When you select Excel's Scenario Manager, you first assign a name to your scenario and list the cells that you wish to change in the scenario. Figure 2.2 shows the next dialog, where you assign the values to those cells that represent the chosen scenario.

We also defined an optimistic scenario, using in this case the values of 0.1 for the consumer sensitivity and 0% for the increased movie attendance, and finally the Base Case scenario using the original values. After defining the two new scenarios, we click on Summary and choose J28 (total monthly profit) as the result cell. The result is shown in Figure 2.3.

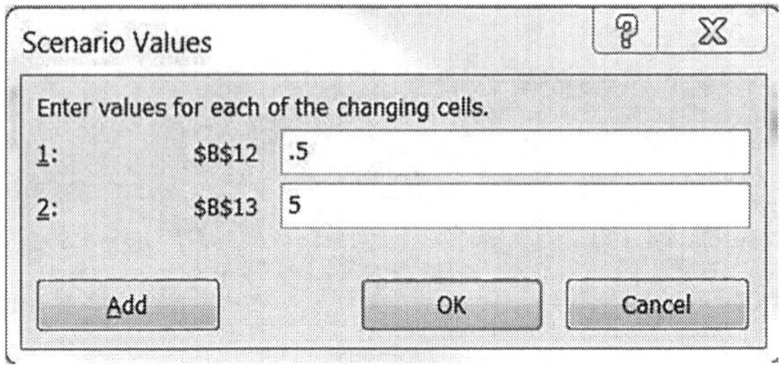

FIGURE 2.2
Pessimistic Scenario Definition

The Summary shows us that our profits for the three scenarios range from −$26,867,627 to +$8,969,940, with the Base Case being −$25,075,534. The Scenario Summary is a concise way to evaluate these scenarios and is particularly useful when you need to evaluate many (i.e., more than three) separate scenarios.

Scenario Summary			
	Current Values:	Pessimistic	Optimistic
Changing Cells:			
$B$12	25%	50%	0%
$B$13	3	5	0.1
Result Cells:			
$J$28	($25,075,534)	($26,867,627)	$8,969,940

Notes: Current Values column represents values of changing cells at time Scenario Summary Report was created. Changing cells for each scenario are highlighted in gray.

FIGURE 2.3
Scenario Summary

However, while it can provide you with some initial ideas about risks and ranges, we caution against using solely the Scenario Manager for analysis purposes.[5] In fact, the Scenario Manager has the danger of appearing to model the uncertainty in the results, without rigorously modeling it. Just how likely are the pessimistic and optimistic scenarios? Are their probabilities equal? Who decided what "pessimistic" meant? Or, if the −$26,867,627 was defined as the "worst-case scenario," does this mean that it is impossible to be worse, or is the probability thought to be negligible (and then what does negligible mean)? Which of the inputs to the analysis are the greatest drivers of the risk in the results? We have seen many uses of the scenario manager in business, but these questions are rarely answered.

In fact, if you try to use the scenario manager more carefully, it may be even less accurate and useful! Suppose you obtained estimates for the pessimistic scenario by asking your marketing experts a precise question such as: For what increase in movie attendance do you believe there is only a 10% chance that it could be that large (or larger)? Similarly, for the consumer sensitivity, you ask for a value that is large enough that the probability of a larger value is only 10%. Then, if the pessimistic scenario in the Scenario Manager uses these two values,[6] the probability of the pessimistic scenario will actually be only 10% × 10% or 1%. Similarly, the optimistic scenario would only have a 1% chance of happening.

To obtain a scenario with only a 10% chance of occurring would require somewhat subtle calculations—and these would become quite complex if there are more than two parameters and some of these are unrelated, while others are related. As a result, the Scenario Summary *appears* to reflect uncertainty about the outcomes but can often be inaccurate when estimating risk, and misleading for decision-making purposes. Uncertainty can easily be overestimated or underestimated (usually it is the latter) using such a crude tool.

2.4 Monte Carlo Simulation

Monte Carlo simulation is a term coined by physicists working in the Los Alamos National Laboratory in the 1940s.[7] The name comes from the similarity to games of chance, for example, blackjack outcomes or roulette wheel spins. While there are many implementations that have been developed, they share the same general structure:

- Specify one or more probability distributions that describe the relevant uncertain variables in a problem.
- Create a random sample from each of these probability distributions.
- Compute an outcome associated with each random sample.
- Aggregate the results of a high number of random samples to describe probabilistically the outcome of the decision problem.

The above four steps can be described in layman's terms as follows: a Monte Carlo model is simulating not just the base case scenarios, but any possible scenario through the use of probability distributions instead of fixed values. Monte Carlo then simulates the model thousands of times in order to get an understanding of the uncertainty around the possible outcomes. In other words, Monte Carlo is sort of doing scenario analysis on steroids and considering actual probabilities.

Computer software is generally used to perform Monte Carlo simulation, but the idea can easily be imagined using paper, pencil, and a coin. Imagine we are interested in estimating the probability of throwing two tails in a row. To do this, we could actually toss two coins in a row many times. In other words, throw two dice, count the number of tails, and record its number. Do this many times (typically thousands) and then look at how many times out of the total tosses (for example, 10,000 "trials") we got two tails.[8] This will give us a very close approximation of the probability, which will be very close to 25% (e.g., with 10,000 trials, we may get 0.2498 or 0.2509 as the fraction of the time we obtained two tails). Fortunately, we don't have to do this "by hand," and we can rely on (Monte Carlo simulation) software to automate this procedure correctly.

2.4.1 Generating Random Numbers

These software packages generate numbers[9] from a uniform distribution, where every number between 0 and 1 has an equal probability of occurring. Since all probability distributions entail cumulative probabilities that lie between 0 and 1, a random number from a uniform distribution can be used to represent the cumulative probability from such a distribution. Then, using a variety of techniques, it is possible to go from the randomly chosen cumulative probability to the underlying value associated with that cumulative probability.[10]

Figure 2.4 illustrates the procedure. The curve represents the cumulative probability distribution for a hypothetical probability distribution measured over some outcomes, x. The randomly chosen number from a uniform distribution gives a number such as y_i,

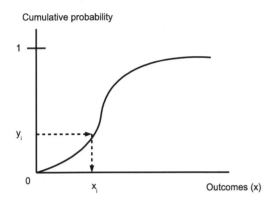

FIGURE 2.4
A Cumulative Probability Distribution

between 0 and 1, which can then be used to compute the value x_i that is associated with that particular number. The techniques for going from a uniform probability to the underlying random outcomes associated with it vary depending on each particular distribution.[11] Also, there are a variety of methods used to generate random numbers from a uniform distribution. Excel uses a particular algorithm, but it is known to not be very robust. Fortunately, most simulation programs (including Monte Carlo Excel add-in packages) use well-tested algorithms that ensure that each number between 0 and 1 is equally probable to be chosen and that the sequence of generated numbers is random.[12]

2.4.2 Monte Carlo Simulation for MoviePass

Let's see how Monte Carlo simulation can be applied to the MoviePass example. We acknowledge that the increase in movie going is unknown, but we may believe it ranges between 10% and 45%, with 25% being the value that we think is most likely. We might also believe that the consumer sensitivity parameter ranges from 1 to 5, with 3 being the most likely value. We will assume these two uncertain variables are independent (we will explore the case where they are correlated in Chapter 5) and both the increased attendance distribution and the consumer sensitivity distribution are distributed according to a Pert distribution (this is a distribution that was developed and is frequently used for modeling expert opinion). The Pert distribution has three input parameters: the minimum, maximum, and most likely values. We explore how distributions may be chosen in Chapter 3. The Pert distributions for both uncertain parameters are illustrated graphically in Figure 2.5.

As you can see, the "peaks" of the two distributions are at the most likely values, 24% and 3%. Further, the min and max values are where the distribution graphs touch zero, illustrating that we assume the values cannot be lower than the min or higher than the max.

Conducting the Monte Carlo simulation then depends on taking random samples simultaneously from these two distributions, computing the outcome for MoviePass monthly profits, repeating this thousands of times, and aggregating the results of all thousands of trials. To accomplish this in an automated way, we turn to the Monte Carlo software packages, such as @Risk.

2.5 Monte Carlo Simulation Using @Risk

In order to follow this example, we assume that you have installed the @Risk software. We are going to replace our two uncertain

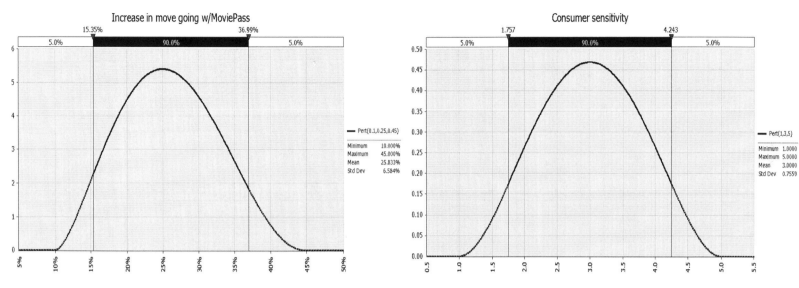

FIGURE 2.5
Pert Distributions for the MoviePass Model

parameters (increased attendance and consumer sensitivity) with the probability distributions shown in Figure 2.5. The @Risk ribbon is shown in Figure 2.6, and we will use the Define Distribution button to enter our distributions.

Selecting the first option in the @Risk ribbon, called Define Distribution, presents us with a list of different sets of distributions (we will discuss these in much more detail in Chapter 3). Scrolling through the Continuous distributions to the Pert and selecting it gives the dialog shown in Figure 2.7.

Regardless of the distribution type, the @Risk distribution window contains a number of consistent features. Along the bottom bar, the user can choose from different views of the probability distribution that the software can show, such as the probability density function[13] or the cumulative probability distribution function. The former shows the probability density of each potential value for the increase in movie attendance, and the latter shows the probability of obtaining a value less than or equal to each of these values. To the right are a number of statistics corresponding to the distributions. The parameters for our distribution (in this case the minimum, mode or most likely, and maximum) are entered on the left side (where we can also restrict the range of values or shift the distribution, using the drop-down menu next to

FIGURE 2.6
The @Risk Menu Bar

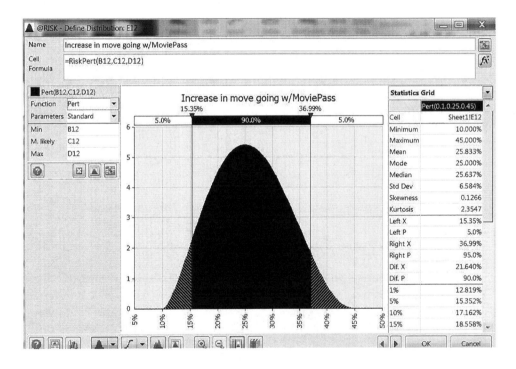

FIGURE 2.7
Entering the Pert Distribution

Basic Monte Carlo Simulation in Spreadsheets

"Parameters"). Note that these values can (and should!) be placed on the spreadsheet in cells and the dialog can be linked to these cells—this is a *much* better modeling technique, since it really keeps the parameters visible on the spreadsheet and is amenable to being easily changed by the user.

When we are done entering the parameters for our increased attendance distribution, we click OK and insert it into the spreadsheet. If we start the Define Distribution with the cursor in cell E12, that is where our input distribution will be placed. We then perform the same steps for the consumer sensitivity parameter in cell E13 of MoviePass.xlsx. After pasting the distribution into cell E13, our spreadsheet formulas are shown in Figure 2.8.

Compare this with Figure 1.2 and you can see that the static values in cells B12 and B13 have been replaced by the minimum, most likely, and maximum values and the Pert distributions using these. It is these distributions, cells E12 and E13, that are used in place of B12 and B13 in the Calculations section of the spreadsheet. In Figure 2.8, the two cells used in the calculations are both functions that start with the word "Risk." The syntax for @Risk functions is that they are functions that begin with the name Risk, followed by what type of function they are, with a number of function arguments. In this case, the functions are Pert distribution functions, and the arguments are the parameters for these distributions that we have assumed.

	A	B	C	D	E
1	Parameters				
2	Attendance distribution		mean (from data)	=SUMPRODUCT(A4:A10,B4:B10)	
3	admissions	fraction			
4	16	0.01	subscription price	9.95	
5	5	0.01	admissions price	8.97	
6	3	0.04			
7	2.1	0.08			
8	0.85	0.19			
9	0.2	0.53			
10	0	0.14			
11		Low	Most likely	High	
12	Increase in move going w/MoviePass	0.1	0.25	0.45	=RiskPert(B12,C12,D12)
13	Consumer sensitivity	1	3	5	=RiskPert(B13,C13,D13)

FIGURE 2.8
MoviePass Model with Uncertainty

Note a few important comments and good practices about these functions:

- They are functions and can be entered and manipulated like other Excel functions. For example, instead of the Pert parameters of a minimum = 60, mode = 80, and maximum = 100, we could (and should) use cell addresses that contain these values.
- As you gain familiarity with these @Risk distribution functions, you may bypass the Select Distribution dialog and directly enter the @Risk function and its arguments into the spreadsheet, just like any other Excel function. These functions are fully supported through help like all other Excel functions.
- These functions can be copied and pasted just like any other Excel function—make sure that you use appropriate absolute or relative cell addressing before copying, however.
- If you want to see the Define Distribution dialog for any reason after you have entered it, simply click Define Distribution (in the @Risk ribbon) when your cursor is on the cell where you entered the distribution and it will bring up the window defining that distribution. You can use the button for "Model Window" (it appears in Figure 2.6) and it will display all the input distributions you have placed in your model (as well as outputs, which we will get to shortly).
- When the spreadsheet values are displayed (not the formulas), you can see the random simulations working by clicking the dice button (the second small button to the right of Settings, near the middle of the @Risk menu bar shown in Figure 2.6) and using F9 (recalculate) on the keyboard. This will show a series of individual random values drawn from the distributions you have placed in the spreadsheet.

To complete our first simulation, we must inform @Risk what outcome(s) we are interested in. For now, let's focus on just the total monthly profit (cell J28). Select cell J28 and click on the Add Output button on the @Risk ribbon. We are now ready to run our Monte Carlo simulations. Click on the Settings button on the @Risk toolbar. Figure 2.9 shows the five buttons we have used in setting up this model: Define Distribution, Add Output, Model Window, the random dice button, and the settings button.

FIGURE 2.9
@Risk Buttons Used Setting Up the Model

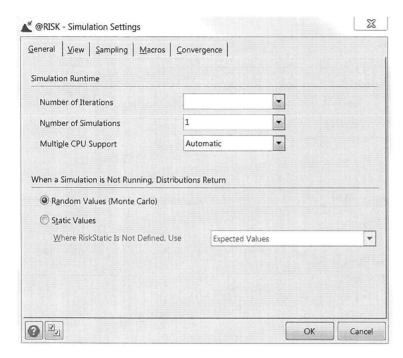

FIGURE 2.10
Simulation Settings Window

Indicate that we will run 1 Simulation, with 10,000 trials (also called "iterations" or "samples"), and return Static Values when the simulation is not running, as shown in Figure 2.10.

Note that there are other options within the Settings window. View will control what you see during the simulation, sampling provides for different types of advanced simulation settings—one to be aware of is the Initial Seed option. If set to Choose Randomly, every time you run a simulation, the results will differ slightly (if you ran a great number of trials, the different will be very similar, though). If you change it to Fixed, then setting an initial seed value (for example, 1) will produce the same results every time you run the simulation (you'd also need to have exactly the same model, of course). Note also that you can change the number of simulations and iterations directly on the menu bar without using the Settings window.[14] TIP: The cell you have highlighted is what will be displayed by default during the simulation, so it makes sense to choose a cell of interest. You can also select a different cell during the simulation, but be careful with this because if you double-click by accident, it may stop the Monte Carlo prematurely.

The simulation is run by clicking the Start button on the @Risk ribbon. After the 10,000 simulations are complete, the results shown in Figure 2.11 should appear.[15]

The default view is a histogram of the 10,000 simulation results with sliders showing a 90% interval around the mean of the total profit distribution (ranging from approximately −$27.5 million to −$18.5 million).[16] The panel on the right side of Figure 2.11 shows a number of statistical summaries, such as the 10th Percentile (−$27,201,757, meaning there is a 10% chance of profits being less than this) or the 90th Percentile (−$20,207,541, for which there is only a 10% chance of profits being greater).

Clicking on the Tornado button (middle bottom of Figure 2.11, with the drop-down arrow) brings up a Tornado chart. This enables understanding the relative importance of the uncertain inputs to the uncertainty in the respective model output. There are only two random inputs here, but from reviewing the Tornado chart, it is clear that consumer sensitivity is far more important than increase in movie attendance for MoviePass profits.

It is also possible to obtain many of the simulation results directly in the spreadsheet, although we recommend also examining

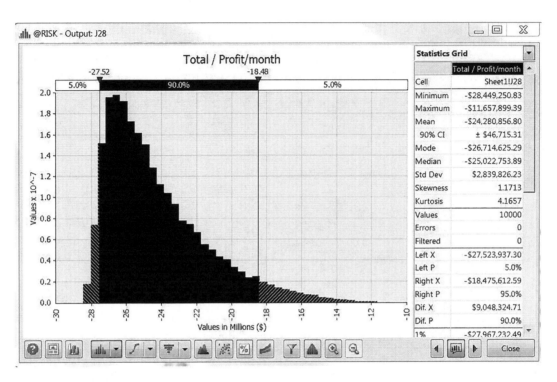

FIGURE 2.11
Initial MoviePass Model Results

the output window for the rich set of features for exploring the outputs. Under Insert Function on the @Risk menu bar, there are quite a few Simulation Result functions that will record a variety of simulation results and place these in the spreadsheet. For example, in MoviePass2a.xlsx, we placed the 5th and 95th percentiles for monthly revenues and profits, along with the mean values from the simulations. Figure 2.12 shows how this is set up in the spreadsheet.

The RiskPercentile function returns a specified percentile (for the 5th and 95th percentiles, we refer to the cells with 0.05 and 0.95, respectively) for a specified cell (in this case, monthly revenues, F28, and monthly profits, J28). Note that when we designated J28 as a

Basic Monte Carlo Simulation in Spreadsheets

	F	G	H	I	J
19	Revenues	Marg. Costs/subscriber	Marg. Profit/subscriber	Total marg. Costs	Profit/month
20	=E20*$D$4	=A20*$D$5	=$D$4-G20	=G20*E20	=F20-I20
21	=E21*$D$4	=A21*$D$5	=$D$4-G21	=G21*E21	=F21-I21
22	=E22*$D$4	=A22*$D$5	=$D$4-G22	=G22*E22	=F22-I22
23	=E23*$D$4	=A23*$D$5	=$D$4-G23	=G23*E23	=F23-I23
24	=E24*$D$4	=A24*$D$5	=$D$4-G24	=G24*E24	=F24-I24
25	=E25*$D$4	=A25*$D$5	=$D$4-G25	=G25*E25	=F25-I25
26	=E26*$D$4	=A26*$D$5	=$D$4-G26	=G26*E26	=F26-I26
27					
28	=SUM(F20:F26)			=SUM(I20:I26)	=RiskOutput()+SUM(J20:J26)
29					
30					
31					
32			0.05		0.95
33		Results:	="Low (P"&H32*100&")"	Mean	="High (P"&J32*100&")"
34		Revenues/mo	=RiskPercentile($F$28,H32)	=RiskMean($F$28)	=RiskPercentile($F$28,J32)
35		Profits/mo	=RiskPercentile($J$28,H32)	=RiskMean($J$28)	=RiskPercentile($J$28,J32)

FIGURE 2.12
Simulation Output Functions in the Spreadsheet

model output, the function RiskOutput was placed in front of that cell's formula calculation.

Astute readers will have noticed that, given the current model inputs, there is no possibility (i.e., a 0% probability) for MoviePass to make a profit. In particular, unless a significant number of the low-attendance consumer segments subscribe, MoviePass will lose money.

So let's look at the Movie Pass business case a bit further. One element of the original business model presented by MoviePass was the expectation that they would obtain bulk discounts on admission prices rather than paying the theater's list price for every attendee. We can investigate this through a simple adaptation to our model (MoviePass2b.xlsx). We consider four potential admissions prices (i.e., costs to MoviePass) that MoviePass might negotiate with theater owners: $6, $7, $8, $9. We place these values in the spreadsheet (cells E5:E8) and then replace the admissions price of $8.97 with the RiskSimTable function (found under Insert Function on the @Risk menu bar) with E5:E8 as the varying values for the table, as shown in Figure 2.13.

The formula for cell E5 is shown, along with a change in the settings to run four simulations rather than just one. We run this simulation (with a fixed seed = 1) and obtain the results shown in Figure 2.14

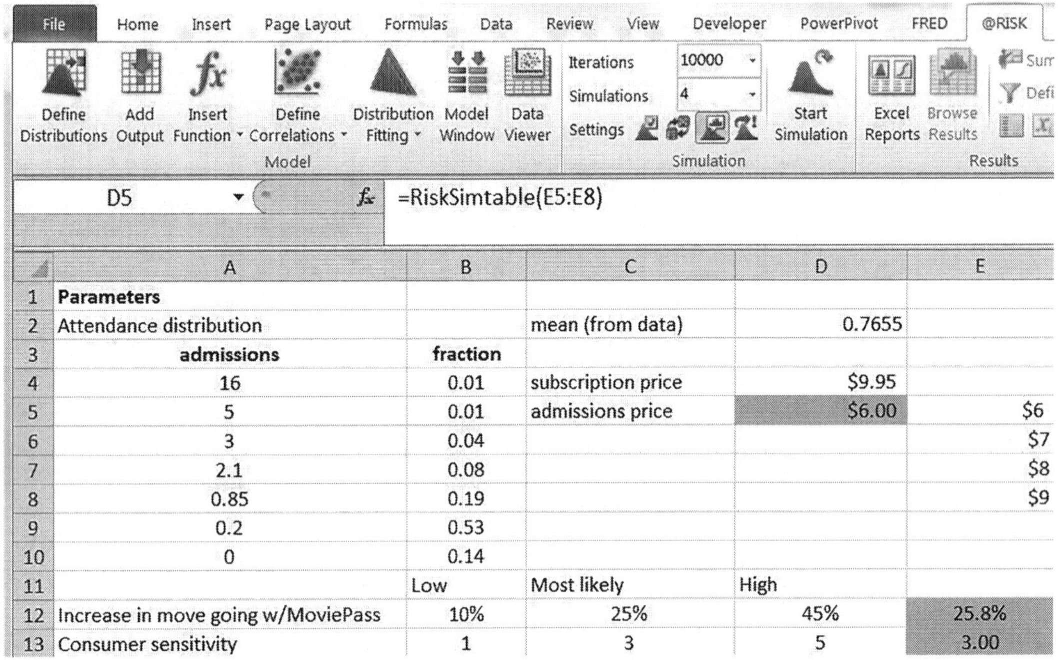

FIGURE 2.13
MoviePass Model with Potential Discounted Admission Prices

	A	B	C	D	E
1	**Parameters**				
2	Attendance distribution		mean (from data)	0.7655	
3	admissions	fraction			
4	16	0.01	subscription price	$9.95	
5	5	0.01	admissions price	$6.00	$6
6	3	0.04			$7
7	2.1	0.08			$8
8	0.85	0.19			$9
9	0.2	0.53			
10	0	0.14			
11		Low	Most likely	High	
12	Increase in move going w/MoviePass	10%	25%	45%	25.8%
13	Consumer sensitivity	1	3	5	3.00

(we overlay the cumulative distributions rather than the histograms since they are clearer).

Profits are clearly higher when we assume that discounted admissions prices will be obtained—but they are still negative, even at the lowest value of $6. The profit distribution is more variable at the higher admissions prices as well. So, while discounted admission prices do improve the profit situation, it is not sufficient to make profitability look likely. Second, consumer sensitivity appears to be the key risk driver—if we change the parameters for the Pert distribution from (1,3,5) to (0,1,2), then profitability looks much more likely, as shown in Figure 2.15.[17]

Figure 2.15 was produced by changing the Graph Options (under Delimiters to show 4 Delimited Curves), moving the sliders to 0 and the maximum, and changing from histograms to the cumulative

Basic Monte Carlo Simulation in Spreadsheets

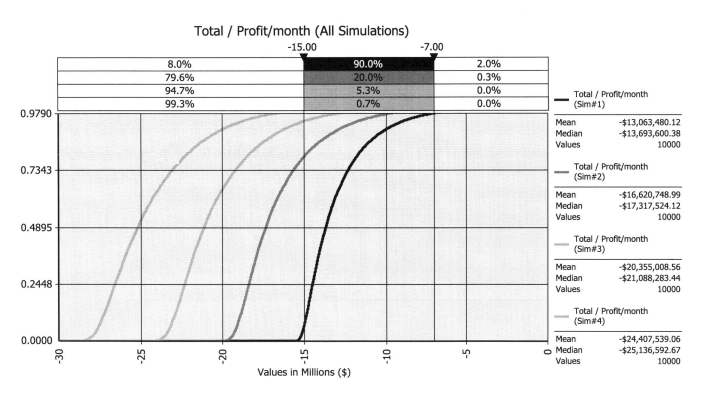

FIGURE 2.14
MoviePass Profit Results with Reduced Admission Prices

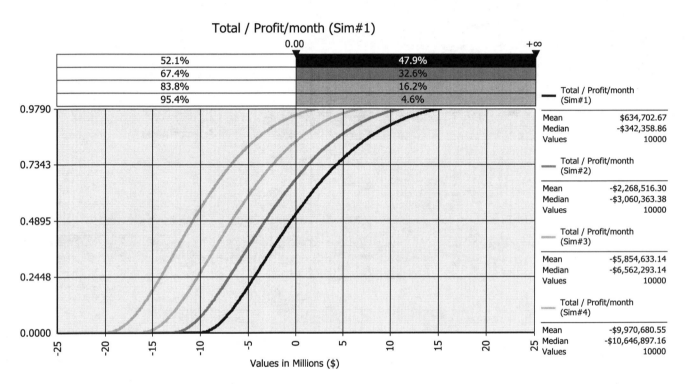

FIGURE 2.15
MoviePass Profit Results with Reduced Consumer Sensitivity

distributions). Now, the probability that MoviePass is profitable ranges from 47.9% with admission prices at $6 to 4.6% with $9 admission prices. This leads to the question of whether/how Movie Pass could get people with a relatively low movie attendance rate to subscribe to their services. Or are they hoping to negotiate even lower admission prices with movie theaters?

2.6 Monte Carlo Simulation for Retirement Planning

We will further explore the use of Monte Carlo simulation by returning to our retirement example from the beginning of this chapter. Figure 2.16 shows our visualization of the retirement

Basic Monte Carlo Simulation in Spreadsheets

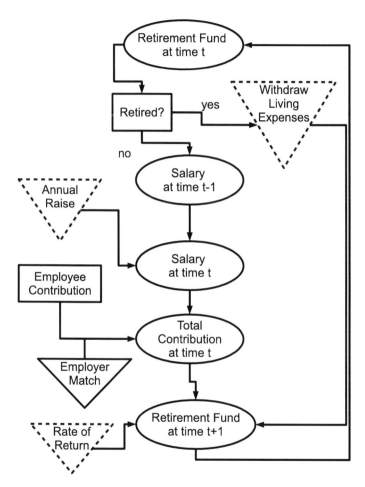

FIGURE 2.16
Retirement Planning Visualization with Uncertainty

planning problem, with three key uncertainties represented: the average annual raise, the average annual living expenses (after retirement), and the annual return for the retirement fund.

Historical stock market data shows that the annual real return of the S&P 500 over the 2000–2017 period was approximately normally distributed with a mean of 4.7% and a standard deviation of 17.5%.[18] We will also assume that the annual raise is uncertain, and we'll use a Pert distribution with a minimum of 0%, a most likely value (mode) of 1%, and a maximum of 3%. Our uncertainty about future average post-retirement living expenses will be assumed to follow a Pert distribution with a minimum of $40,000, mode of $50,000, and maximum of $70,000 (note: all figures are assumed to be adjusted for inflation). This uncertainty is supposed to reflect uncertainty about what our individual's future lifestyle and health, and thus spending level, might look like. For example, one reason for the potentially larger annual expenses may be that their health status declines and they incur increased out-of-pocket health care costs. Finally, we assume that their retirement age is expected to be 65.[19]

We show these assumptions in Figure 2.17 (cell B7) but must make an additional modification to our spreadsheet. The annual return on the retirement fund is uncertain, but we could not expect each year's return to be the same. So, we can't just simulate one year's return and use that in our model for each future year—instead, we must simulate each year's return. We do this by modifying Column D of the spreadsheet to calculate the return by multiplying the fund size in Column C of the spreadsheet by a function, RiskNormal(C7, D7). Every year will now have its own simulated asset return.

Figure 2.17 shows the distributions we have entered along with the modifications to our previous retirement model. The annual rate

	A	B	C	D	E	F	G	H	I
1	Parameters		minimum	mode	maximum	Simulation #	Contribution %	mean yr runs out	Prob lasts > 81 age
2	Current Age	30				1	0.05	=RiskMean($C$11,F2)	=1-RiskTarget($C$11,81,F2)
3	Salary	50000				2	0.06	=RiskMean($C$11,F3)	=1-RiskTarget($C$11,81,F3)
4	Annual Raise	=RiskPert(C4,D4,E4)	0	0.01	0.03	3	0.07	=RiskMean($C$11,F4)	=1-RiskTarget($C$11,81,F4)
5	Employee Contribution	=RiskSimtable(G2:G12)				4	0.08	=RiskMean($C$11,F5)	=1-RiskTarget($C$11,81,F5)
6	Employer Match	0.07		(mean)	(standard deviation)	5	0.09	=RiskMean($C$11,F6)	=1-RiskTarget($C$11,81,F6)
7	Rate of Return	=RiskNormal(C7,D7)	0.047	0.175		6	0.1	=RiskMean($C$11,F7)	=1-RiskTarget($C$11,81,F7)
8	Retirement Living Expenses	=RiskPert(C8,D8,E8)	40000	50000	60000	7	0.11	=RiskMean($C$11,F8)	=1-RiskTarget($C$11,81,F8)
9	Initial Fund Balance	0				8	0.12	=RiskMean($C$11,F9)	=1-RiskTarget($C$11,81,F9)
10	Decisions		Objective (year funds run out)			9	0.13	=RiskMean($C$11,F10)	=1-RiskTarget($C$11,81,F10)
11	Retirement Age	65	=RiskOutput()+B2+SUM(L15:L85)			10	0.14	=RiskMean($C$11,F11)	=1-RiskTarget($C$11,81,F11)
12						11	0.15	=RiskMean($C$11,F12)	=1-RiskTarget($C$11,81,F12)
13	Calculations			average annual return	=AVERAGE(D15:D49)				
14	Age	Retired?	Fund at Start of Year	Annual Return %	Return on Fund	Salary	Employee Contribution	Employer Match	Total Contributions
15	=B2	=IF(A15>=$B$11,1,0)	=$B$9	=RiskNormal($C$7,$D$7)	=C15*D15	=(1-B15)*($B$3*(1+$B$4)^(A15-30))	=$B$5*F15	=IF($B$5>7%,7%,$B$5)*F15	=G15+H15
16	=A15+1	=IF(A16>=$B$11,1,0)	=K15	=RiskNormal($C$7,$D$7)	=C16*D16	=(1-B16)*($B$3*(1+$B$4)^(A16-30))	=$B$5*F16	=IF($B$5>7%,7%,$B$5)*F16	=G16+H16
17	=A16+1	=IF(A17>=$B$11,1,0)	=K16	=RiskNormal($C$7,$D$7)	=C17*D17	=(1-B17)*($B$3*(1+$B$4)^(A17-30))	=$B$5*F17	=IF($B$5>7%,7%,$B$5)*F17	=G17+H17
18	=A17+1	=IF(A18>=$B$11,1,0)	=K17	=RiskNormal($C$7,$D$7)	=C18*D18	=(1-B18)*($B$3*(1+$B$4)^(A18-30))	=$B$5*F18	=IF($B$5>7%,7%,$B$5)*F18	=G18+H18
19	=A18+1	=IF(A19>=$B$11,1,0)	=K18	=RiskNormal($C$7,$D$7)	=C19*D19	=(1-B19)*($B$3*(1+$B$4)^(A19-30))	=$B$5*F19	=IF($B$5>7%,7%,$B$5)*F19	=G19+H19
20	=A19+1	=IF(A20>=$B$11,1,0)	=K19	=RiskNormal($C$7,$D$7)	=C20*D20	=(1-B20)*($B$3*(1+$B$4)^(A20-30))	=$B$5*F20	=IF($B$5>7%,7%,$B$5)*F20	=G20+H20
21	=A20+1	=IF(A21>=$B$11,1,0)	=K20	=RiskNormal($C$7,$D$7)	=C21*D21	=(1-B21)*($B$3*(1+$B$4)^(A21-30))	=$B$5*F21	=IF($B$5>7%,7%,$B$5)*F21	=G21+H21
22	=A21+1	=IF(A22>=$B$11,1,0)	=K21	=RiskNormal($C$7,$D$7)	=C22*D22	=(1-B22)*($B$3*(1+$B$4)^(A22-30))	=$B$5*F22	=IF($B$5>7%,7%,$B$5)*F22	=G22+H22
23	=A22+1	=IF(A23>=$B$11,1,0)	=K22	=RiskNormal($C$7,$D$7)	=C23*D23	=(1-B23)*($B$3*(1+$B$4)^(A23-30))	=$B$5*F23	=IF($B$5>7%,7%,$B$5)*F23	=G23+H23
24	=A23+1	=IF(A24>=$B$11,1,0)	=K23	=RiskNormal($C$7,$D$7)	=C24*D24	=(1-B24)*($B$3*(1+$B$4)^(A24-30))	=$B$5*F24	=IF($B$5>7%,7%,$B$5)*F24	=G24+H24
25	=A24+1	=IF(A25>=$B$11,1,0)	=K24	=RiskNormal($C$7,$D$7)	=C25*D25	=(1-B25)*($B$3*(1+$B$4)^(A25-30))	=$B$5*F25	=IF($B$5>7%,7%,$B$5)*F25	=G25+H25
26	=A25+1	=IF(A26>=$B$11,1,0)	=K25	=RiskNormal($C$7,$D$7)	=C26*D26	=(1-B26)*($B$3*(1+$B$4)^(A26-30))	=$B$5*F26	=IF($B$5>7%,7%,$B$5)*F26	=G26+H26
27	=A26+1	=IF(A27>=$B$11,1,0)	=K26	=RiskNormal($C$7,$D$7)	=C27*D27	=(1-B27)*($B$3*(1+$B$4)^(A27-30))	=$B$5*F27	=IF($B$5>7%,7%,$B$5)*F27	=G27+H27
28	=A27+1	=IF(A28>=$B$11,1,0)	=K27	=RiskNormal($C$7,$D$7)	=C28*D28	=(1-B28)*($B$3*(1+$B$4)^(A28-30))	=$B$5*F28	=IF($B$5>7%,7%,$B$5)*F28	=G28+H28

FIGURE 2.17
Retirement Model Including Uncertainty

of return is simulated in Column D and the arithmetic average of these returns (during the working life) is calculated in cell D13. Note that the probability distributions refer to cell addresses holding the parameters of these distributions. We have also set up a section of the spreadsheet (G1:I12) to explore the effects of varying the employee contribution to the retirement fund. This is the one variable the individual has the most control over, and a major reason for modeling retirement funds is to provide information to help plan contributions to retirement plans.

We are now ready to run the simulation. Again, if you want to match our exact numerical outputs, set the fixed seed to 1, and run 11 simulations of 10,000 iterations in order to get the same results. Our initial results window (showing simulation #1) is shown in Figure 2.18.

Immediately noticeable in Figure 2.18 is the tall bar at age 100. We only built our model to go to age 100, and in around 10% of the simulations, the retirement fund was not exhausted at that age. This is certainly encouraging for our retirement plans, but the fact that

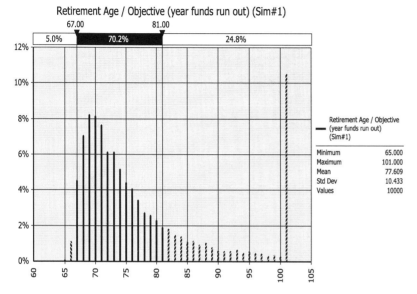

FIGURE 2.18
Results for Contribution of 5%

almost 75% of the time the fund does not last until an age of 81 years (which is close to our life expectancy) is worrisome (move the slider position on the right to 81 to see 24.8% in the results histogram). Further insight is available through reviewing the tornado chart. Click on the Tornado chart display type at the bottom of the Simulation Results window (note that you can change the output for any of the simulation runs by using the third button from the right along the bottom of the output window) to get Figure 2.19.

The tornado chart shows the relative importance of our uncertain parameters. The most significant factors are the average annual returns, and it is the returns near retirement age that affect the variability of when funds run out the most (returns near retirement age are known key risk factors for retirement planners, and this simple @Risk model confirms this well).[20]

To see how the employee contribution affects the results, we can use the results window and click on the button for overlaying results. Select our "year funds run out" cell (C11) and we get a dialog that asks which simulation runs we wish to overlay. Choose all (although only 10 will be displayed). The overlaid histograms are too difficult to read, so click the Summary Trend button, once again select cell C11 and all simulations, and then use the button at the bottom to change the display from a trend chart to box plots. Figure 2.20 shows the result.

These box plots show the middle 50% of each simulation as a box (middle two quartiles), with the two lines extending from the box representing the lowest and highest 25% of the simulations (i.e., the low and high quartile). This box plot is also known as a "box and whisker plot." Figure 2.20 clearly shows that the more the employee contributes, the longer their retirement fund is expected to last (the mean is the expected age of running out of funds). The variability of the results is a bit harder to discern. We get a clearer view of these results by using the section of the spreadsheet we set up for this purpose, cells F1:I12, shown in Figure 2.21.

Column I displays the probability that the funds last at least until an age of 81 years and shows that this probability increases (though not linearly) as we increase our contribution level. However, even with an annual contribution of 15%, there is still a 35.9% chance we will run out of funds by age 81. Thus, if we find these results too risky, then we would need to explore further changes, such as planning to reduce our post-retirement expenses, raise our retirement age, or even raise our annual retirement contribution level further, beyond the 15%.

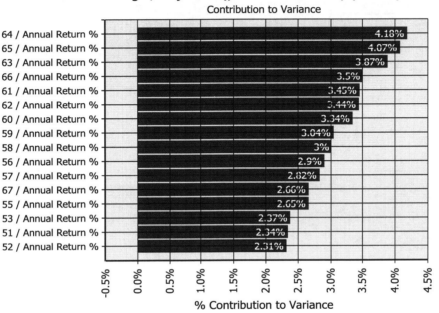

FIGURE 2.19
Tornado Chart for the Retirement Model

2.7 Discrete Event Simulation

This completes our initial foray into Monte Carlo simulation. We have seen how to set up and interpret a simulation, including the use of tornado charts and simulation tables. The next chapter will explore how we can choose appropriate probability distributions, including fitting these to data.

We end this chapter with a note about types of simulation models that may not be amenable to Monte Carlo simulation within spreadsheets. When you are interested in modeling a project where it is important to simulate the chronological sequence of events (for example, ship movements in a harbor or patients' stays in a hospital), typically this can best be modeled using a technique called discrete event simulation (DES). Software packages that offer DES capabilities generally have a "clock," an event list, and a random-number

Basic Monte Carlo Simulation in Spreadsheets

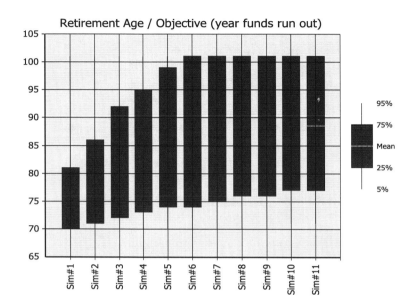

FIGURE 2.20
Year When Funds Run Out as a Function of the Contribution Level

	F	G	H	I
1	Simulation #	Contribution %	mean yr runs out	Prob lasts > 81 age
2	1	5%	77.6	25%
3	2	6%	80.0	33%
4	3	7%	82.2	41%
5	4	8%	83.1	44%
6	5	9%	84.0	47%
7	6	10%	84.9	51%
8	7	11%	85.8	54%
9	8	12%	86.5	57%
10	9	13%	87.3	59%
11	10	14%	87.9	62%
12	11	15%	88.6	64%

FIGURE 2.21
Spreadsheet Results for Varying Contribution Levels

generator. The general idea of discrete event simulation is that a clock starts and, after a random amount of time, events will happen that will have particular effects. For example, a simple problem would be a queue at a bank where customers randomly arrive, and it takes a random amount of time to help each customer. Similarly, patients arrive randomly in a hospital, with different diseases and severity, and random needs for treatment but possibly a constrained number of beds.

While some simple DES models can be modeled within Excel (possibly with some VBA), modeling such problems in Excel is typically more difficult since it doesn't have a convenient time tracker nor an easy way keeping track of all the possible events. For fairly simple discrete event problems, some Excel logic (IF statements and so on) can be used, but once there are several different events possible, implementing this in Excel quickly becomes too cumbersome. Dedicated software packages such as Simul8, Anylogic, or Arena are in such cases much better and more efficient choices.

One notable exception is when using project planning software, such as Microsoft Project. This software keeps track of the timing at which different parts of a project occur, as well as the relationships between different tasks in a project. Furthermore, @Risk is designed to work with Microsoft Project to enable uncertainty to be built into project planning. The Excel (with @Risk) spreadsheet contains the uncertain inputs and resulting outputs, but Microsoft Project rebuilds the project plan for every iteration that reflects the underlying uncertainty. This application of @Risk is covered in the Appendix of this book.

Exercises

These exercises are a continuation of those from Chapter 1 (in the same order). Starting with your base case model, you are to incorporate the uncertainty described in each exercise and conduct an appropriate Monte Carlo simulation.

Exercise 2.1 Health Insurance Choice Simulation

Use your health care insurance plan choice model from Exercise 1.1. Health care service utilization varies across individuals for myriad reasons. For each family member, assume the actual number of office visits follows a Poisson distribution, with a mean rate (designated by the one input parameter to the Poisson distribution, which is called lambda, λ). While Table 1.2 shows average inpatient costs, in reality these are highly variable. Assume that these follow lognormal distributions, with means and standard deviations both equal to the Table 1.2 values (this will produce highly skewed distributions). In the United States, health care premiums are tax exempt (that is, these premiums are not subject to income taxes—assume a marginal tax rate of 35%). Simulate the cost of the three health care plan options from Exercise 1.1, recognizing the tax benefits and without including the tax benefits. Estimate the probability of our hypothetical family subscribing to each plan, based on their expected health care costs, both with and without tax benefits (this can also be interpreted as the proportion of families that will subscribe to each plan). Compare the variability of the costs under each plan for our family, and consider how this might affect their choice of plan. Write a paragraph describing your findings.

Exercise 2.2 Electric Vehicle Cost Simulation

Use your electric vehicle model from Exercise 1.2. Gasoline and electricity prices are highly uncertain. Investigate how this uncertainty impacts the breakeven additional price for the electric engine. Assume that annual change in gasoline price (adjusted for inflation) follows a normal distribution (mean = −0.2%, standard deviation = 13.4%). Annual changes in electricity prices follow a normal distribution (mean = −0.5%, standard deviation = 2.7%).[21] What is the mean breakeven price for the electric engine? What is the probability that the electric engine is more expensive in terms of net present value?

Exercise 2.3 Educational Investment Simulation

Use your model for estimating the NPV from Exercise 1.3. One source of uncertainty is the possibility of becoming unemployed. High school graduates have an unemployment rate of 5.3%, while those with college degrees (or higher) average 2.5% (use a Bernoulli distribution to model whether a person is employed). Assume that each year these represent the probability of being unemployed, and that a spell of unemployment lasts exactly 1 year. Also assume that each year's chance of being unemployed is independent of all other years, and ignore unemployment compensation in this exercise.

As further sources of uncertainty, consider the starting salary to have a normal distribution with the means given in Exercise 1.3 and standard deviations equal to 20% of the means. Simulate NPV for the two investment decisions, at two discount rates, 5% and 10% (these rates are indicative of the difference between student federal loans and private student loans). What is the probability that the NPV is positive for each investment (at each of these discount rates)?

Exercise 2.4 Customer Lifetime Value Simulation

Use your LTV model from Exercise 1.4. One source of uncertainty was already presented in Exercise 1.4—the retention rate. Rather than assuming that the average retention rate is experienced with certainty, simulate the actual retention of customers (use the binomial distribution for this). Also, assume that the actual retention rate is not known precisely, but follows a PERT distribution with uncertainty of +/−5% around each of these rates. Provide 90% confidence intervals for the LTV after 1, 2, and 3 years.

Exercise 2.5 Network Economics Simulation

Consider your network economics model from Exercise 1.5. There are two major sources of uncertainty that should be accounted for. For a new communications technology, the costs will initially not be known with certainty. Assume that the costs may range between $3 and $8 (with a most likely value of $5, following a PERT distribution). Also, the intrinsic valuations can only be estimated from market research. Assume that the intrinsic valuation function $(100(1 - F))$ provides an expected estimate, but that the real valuations must be multiplied by an uncertainty factor. This factor will follow a normal distribution with a mean of 1 and a standard deviation of 0.2.

Determine 90% confidence intervals for the critical mass and saturation levels.

Exercise 2.6 Peak Load Pricing Simulation

In Exercise 1.6, we saw that peak load pricing can increase profitability by leveling use between peak and off-peak periods (thereby reducing the need for expensive capacity). In this exercise, we examine how uncertain demand influences our results. Suppose that the peak and off-peak demand levels from Exercise 1.6 represent the average hourly demands during these times. Represent the uncertainty by multiplying the average demand by a normal distribution (with mean = 1, standard deviation = 0.25). Each hourly demand will be random, and assume that each varies independently of the other hours. Compare the average daily profitability for the single price (50) scheme and the peak-load pricing scheme (75,25). What is the probability that peak load pricing will be more profitable than the single price scheme? Produce a graph that shows how hourly demand levels fluctuate during the day under the two pricing schemes.

Exercise 2.7 Project Management Simulation

Use the project management model from Exercise 1.7. Projects are notorious for running over budget and taking longer than anticipated. Suppose that each activity time from Table 1.4 is uncertain and follows a PERT distribution with the table numbers being the most likely values, but with minimum values = 50% of the table entries and maximum values = 150% of the table entries (there are typically more things that cause a task to take more time rather than less, but the results will be more insightful if we assume symmetric uncertainty). Calculate the expected time to project completion and provide a 90% confidence interval for the time to completion. What is the probability that the time will exceed 15 months (i.e., the answer from Exercise 1.7)?

Exercise 2.8 Hollywood Finance Simulation

Use your model for the profitability of an average Hollywood movie from Exercise 1.8. The variability of movie profitability has received much study and is the subject of plenty of casual conversation.[22] Suppose the costs (production plus marketing) are uncertain and follow a PERT distribution (minimum = $50 million, mode = $100 million, maximum = $450 million). Movie production can take anywhere from 6 to 24 months (following a uniform distribution). Domestic box office receipts also follow a PERT distribution (minimum = $0, mode = $0, maximum = $200 million—note that this assumption reflects the fact that many movies never run in theatres at all). Home video sales are still based on domestic box office receipts, but follow a PERT distribution (minimum = 20%, mode = 80%, maximum = 180%). Hit movies (defined as those with greater than $100 million in domestic box office receipts) earn $25 million from premium TV and $25 million from free TV—higher than the figures used in Exercise 1.8.

Estimate the internal rate of return for films.[23] Use the Excel function XIRR (for monthly rates of return).[24] Note that the IRR function will produce Excel errors if the true internal rate of return is a large negative number. The XIRR function also encounters problems. These are related to the initial guess (the third, and optional, argument in the function). The default guess is 0.1, and this will produce errors if the true internal rate of return is negative and large enough (in absolute value). To remedy this problem, use an initial guess equal to 0.1 * sign(sum of cash flows). When the sum of the cash flows is negative (positive), this results in an initial guess of −0.1 (+0.1). XIRR will then correctly estimate the internal rate of return for both negative and positive values.

What fraction of films earn a negative return? What fraction earns returns above 30%? (You can compare these fractions to those identified in the footnote. They will not be the same, but they should be qualitatively similar).

Notes

1. Knight, F. 1921. *Risk, Uncertainty and Profit*. Hart, Schaffner, and Marx.
2. Taleb, N. 2007. *The Black Swan: The Impact of the Highly Improbable*. Random House.
3. Marr, B. 2018. "How Much Data Do We Create Every Day? The Mind-Blowing Stats Everyone Should Read." *Forbes*.
4. Berra, L. P. 2002. *When You Come to a Fork in the Road, Take It!: Inspiration and Wisdom from One of Baseball's Greatest Heroes*. Hyperion.
5. When evaluating the outcomes of different discrete decisions, the scenario manager may be particularly useful.
6. And if we assume our estimates of the two parameters are independent.
7. Metropolis, N. and Ulam, S. 1949. "The Monte Carlo Method." *Journal of the American Statistical Association*, 44: 335–341.
8. Strictly speaking, Monte Carlo simulation becomes increasingly accurate as the number of times the procedure is repeated approaches infinity (Monte Carlo methods are based on the Law of Large Numbers). In practice, we are typically satisfied after 10,000 iterations (in some situations, particularly when we are interested in very low-probability events, more iterations may be required for precision).
9. This subsection is fairly technical and not necessary for understanding the rest of this chapter.
10. Don't worry if you did not fully understand how exactly software packages can generate random numbers. What is important is that software packages can generate random numbers from many different distributions, and therefore greatly facilitate Monte Carlo simulation.
11. A good example for the triangular distribution can be found in Evans, J. R. and Olson, D. L. 2002. *Introduction to Simulation and Risk Analysis*. Prentice Hall.
12. An easy way to think of random number generator is to take an eight-digit number (with a decimal point in front) and select the middle four digits (the random seed). Then, square this number—the result

is another eight-digit number. The procedure is continually repeated, with the resulting numbers being randomly distributed between 0 and 1. While this works most of the time, problems can emerge. For example, if the middle four digits produce .0001, then this procedure will get stuck at .0001 which is clearly not a random sequence of numbers. This algorithm would not be *robust*. Random number generators, such as the Mersenne Twister random number generator that is used in the @Risk software, have been tested and documented for robustness to avoid pitfalls such as these.

13. The probability density of a continuous distribution is a somewhat difficult concept since it is not the same as a probability. However, similar to a probability, the higher the probability density, the higher the likelihood of a value within that corresponding region. The difference is that when we have a continuous distribution, we can only talk about the probability of being *between* two values, or above or below a certain value, and *not* about the probability of exactly a certain value. The probability density values themselves are therefore typically not of great interest.

14. We will discuss the number of simulations later—further "what if" analyses can be obtained that vary parameters within the model without simulating uncertainty for these. An example might be to vary the subscription price to a number of potential values, running a simulation (with 10,000 iterations) for each of these values.

15. If you want to match our numerical results, set the random seed to a fixed value = 1.

16. You can change the 90% interval by either dragging the sliders or using the settings under the Graph Options button below the histogram. A number of additional options are available, which will be covered as we move along.

17. MoviePass2c.xlsx contains the revised model.

18. In Chapter 3, we will show how this can be derived from historical data and how to find the appropriate distribution to use. A normal distribution does not actually fit the data best—the best fitting distribution has fatter tails, which is an important consideration for retirement planning since this means that extreme outcomes are more probable than the normal distribution permits. For now, we will use the normal distribution, as it is probably more familiar to most readers. Retirement2.xlsx contains the data from 1871–2017, and we will reexamine the data for the whole time period in Chapter 3. Note that the return data includes both stock price appreciation and dividends and is adjusted for inflation.

19. Further enhancements to the model would likely explore the retirement age more fully. It is partially a decision, partially random, and partially an objective. For instance, a person might want to minimize the retirement age subject to constraints on the size of the retirement fund (an objective). Alternatively, they might be choosing the retirement age (a decision) in order to achieve particular goals of retirement lifestyle and length. To some extent, their retirement age may be beyond their control (uncertain) as it may be affected by their health or by obsolescence of their workplace skills.

20. For this display, we have used the drop-down menu for the tornado chart and chosen Contribution to Variance.

21. The annual change data for gasoline and electricity prices approximate the US data for the 1976–2015 time frame.

22. See Ferrari, M. J. and Rudd, A. 2008. "Investing in Movies." *Journal of Asset Management*, 19, 1 for some data. They estimate that 62% of films earn a negative return and 26.6% earn over a 30% return on investment—the average is 3%.

23. Assume that there is no relationship between the cost of making a film and its box office revenues—in reality, these are correlated. Correlation is the subject of Chapter 4.

24. The XIRR function requires a date column as an input, as well as a series of values (with the date column in a recognizable Excel date format). The XIRR function avoids some issues that arise when using the IRR function, which produces errors if the rate of return is too negative (as it is for many films). However, both XIRR and IRR have other issues, as described in Exercise 1.8.

3

Selecting Distributions

Learning Objectives

- Understand how distributions may be selected based on theoretical considerations and/or expert opinion.
- Learn how to incorporate expert opinions into spreadsheet models. Apply these methods to a company valuation.
- See how distributions may be fit to historical data.
- Learn ways to estimate and incorporate uncertainty about the appropriate distribution that fits the data. Apply these methods to estimation of the value at risk of a hypothetical portfolio of assets.
- Understand a variety of classes of distributions commonly used to model different types of situations with uncertainty or variability.
- Adopt a structured approach to selecting appropriate probability distributions.

3.1 First Introductory Case: Valuation of a Public Company Using Expert Opinion

Facebook, the world's largest online social network, had more than 2 billion monthly active users (more than 25% of the world's population) at the end of 2018. It was also immensely profitable, with gross margins above 80%. Advertising revenues make up more than 90% of the company's total revenue. One way to value a company is through a discounted cash flow analysis, in which the worth of a company is estimated by totaling all expected future cash flows and discounting them back to the current date. To do this, the net cash flows (a.k.a. free cash flows, FCFs) are projected for a period of time (typically 5–10 years) and a terminal value is added to reflect ongoing cash flows after that period (usually at a fairly low level of sales growth, or multiple of sales). Discounting is commonly done at the weighted average cost of capital (WACC), which can be seen as a weighted average of the firm's debt and equity. The cost of equity is typically calculated using the capital asset pricing model (CAPM), where the cost of equity includes three components: a riskless interest rate, an equity premium reflecting the riskiness of equity markets overall, and a company-specific additional risk factor (beta, reflecting the riskiness[1] of a company compared with the market overall). Figure 3.1 shows the most recent 5 years of income statement data for Facebook (sticker FB traded on NASDAQ).

In this example, the assumed forecasts are for 20% annual revenue growth for the next 5 years.[2] Terminal revenue growth (from year 10 onward) is assumed to be 3%, similar to the historical long-term growth of the world economy. To avoid a sudden decrease in revenue growth after year 10, revenue growth is linearly decreased after year 5 to equal the terminal growth rate in year 10. In addition to the future revenue forecasts, several other key ratios are important for the projected income and cash flow statement. In the initial base case situation, most of the ratios (cost of goods sold, R&D expenditures, and SG&A expenditures) are assumed to be simply the average of the ratios over the past 5 years. Also, the tax rate is assumed to be the average from the past 5 years. Of course, all of these assumptions about the future financial performance of Facebook can be, and should be, questioned. Indeed, as you'll see in the next section, a principal purpose of this model is to explicitly consider the major uncertainties and examine their impact on FB's valuation. Finally, the cost of equity[3] is estimated to be 9.8%, using a riskless rate of 3%, a market risk premium of 6%, and a company beta of 1.14 (estimated from stock market data).[4] Based on these inputs, the valuation of Facebook comes to over $418 billion, as shown in the spreadsheet Valuation3a.xlsx, and in Figure 3.2 (only the first 5 forecast years are shown).

The results of these formulae are shown in Figure 3.3.

While this model's valuation is reasonably close to the market beliefs of FB's value in early 2019 (the market capitalization of Facebook was $428.22 billion on January 28, 2019), it does not consider the myriad uncertainties that the company faces.[5] The use of Monte Carlo simulation to take into account a range of expert opinions can provide a sense of the range of plausible future financial performances (and thus valuations) for FB.

	A	B	C	D	E	F
1	FACEBOOK INC A (FB) CashFlowFlag INCOME STATEMENT					
2	Fiscal year ends in December. USD in millions.	2013-12	2014-12	2015-12	2016-12	2017-12
3	Revenue		7872	12466	17928	27638	40653
4	Cost of revenue		1875	2153	2867	3789	5454
5	Gross profit		5997	10313	15061	23849	35199
6	Research and development		1415	2666	4816	5919	7754
7	Sales, General and administrative		1778	2653	4020	5503	7242
8	Total operating expenses		3193	5319	8836	11422	14996
9	Operating income		2804	4994	6225	12427	20203
10	Interest Expense		56	23	23	10	6
11	Other income (expense)		6	-61	-8	101	397
12	Income before taxes		2754	4910	6194	12518	20594
13	Provision for income taxes		1254	1970	2506	2301	4660
14	Net income from continuing operations		1500	2940	3688	10217	15934
15	Net income		1500	2940	3688	10217	15934

FIGURE 3.1
Facebook Income Statements

3.2 Modeling Expert Opinion in the Valuation Model

The people who develop and run risk analysis models are often not the people with the best insight concerning the range and probability of critical forecasts used in that analysis. Further, any one individual is likely to be wrong in their predictions, while use of several divergent opinions may produce more reliable estimates.[6] For the Facebook valuation, we have identified four key uncertain variables in our analysis: the cost of goods sold (COGS) and R&D

Selecting Distributions

	A	B	C	D	E	F	G	H	I	J	K	L
1	FACEBOOK INC A (FB) CashFlowFlag INCOME STATEMENT						Forecasts					
2	Fiscal year ends in December. USD in millions.	2013-12	2014-12	2015-12	2016-12	2017-12	2018	2019	2020	2021	2022	2023
3	Revenue	7872	12466	17928	27638	40653	=F3*(1+G17)	=G3*(1+H17)	=H3*(1+I17)	=I3*(1+J17)	=J3*(1+K17)	=K3*(1+L17)
4	Cost of revenue	1875	2153	2867	3789	5454	=$G$18*G3	=$G$18*H3	=$G$18*I3	=$G$18*J3	=$G$18*K3	=$G$18*L3
5	Gross profit	5997	10313	15061	23849	35199						
6	Research and development	1415	2666	4816	5919	7754	=$G$19*G3	=$G$19*H3	=$G$19*I3	=$G$19*J3	=$G$19*K3	=$G$19*L3
7	Sales, General and administrative	1778	2653	4020	5503	7242	=$G$20*G3	=$G$20*H3	=$G$20*I3	=$G$20*J3	=$G$20*K3	=$G$20*L3
8	Total operating expenses	3193	5319	8836	11422	14996						
9	Operating income	2804	4994	6225	12427	20203						
10	Interest Expense	56	23	23	10	6						
11	Other income (expense)	6	-61	-8	101	397						
12	Income before taxes	2754	4910	6194	12518	20594	=G3-G4-G6-G7	=H3-H4-H6-H7	=I3-I4-I6-I7	=J3-J4-J6-J7	=K3-K4-K6-K7	=L3-L4-L6-L7
13	Provision for income taxes	1254	1970	2506	2301	4660	=$G$21*G12	=$G$21*H12	=$G$21*I12	=$G$21*J12	=$G$21*K12	=$G$21*L12
14	Net income from continuing operations	1500	2940	3688	10217	15934						
15	Net income	1500	2940	3688	10217	15934	=G12-G13	=H12-H13	=I12-I13	=J12-J13	=K12-K13	=L12-L13
16	Key Ratios											
17	Revenue growth		=(C3-B3)/B3	=(D3-C3)/C3	=(E3-D3)/D3	=(F3-E3)/E3	-0.2	=G17	=H17	=I17	=J17	=K17-($K$17-$B$23)/5
18	COGS	=B4/B3	=C4/C3	=D4/D3	=E4/E3	=F4/F3	=AVERAGE(B18:F18)					
19	R&D	=B6/B3	=C6/C3	=D6/D3	=E6/E3	=F6/F3	=AVERAGE(B19:F19)	averages				
20	SG&A	=B7/B3	=C7/C3	=D7/D3	=E7/E3	=F7/F3	=AVERAGE(B20:F20)					
21	income tax rate	=B13/312	=C13/C12	=D13/D12	=E13/E12	=F13/F12	=AVERAGE(B21:F21)					
22												
23	Teriminal Growth Rate	0.03										
24	Beta	1.14										
25	Riskless Rate	0.03										
26	Risk Premium	0.0596										
27	Cost of capital	=B25+B24*B26										
28	PV net income (10 years)	=NPV(B27,G15:P15)										
29	Terminal Value	=P15/(B27-B23)										
30	PV of terminal value	=B29/((1+B27)^10)										
31	Total Value	=B28+B30										

FIGURE 3.2
Initial Facebook Valuation Model

costs (as a percentage of revenues), the revenue growth rate, and the estimate of beta (which directly affects the WACC and discount rate). Facebook is unusual in that its "production costs" are quite low (many industries show a COGS of 75% or larger). Since Facebook does not produce any tangible product, this is not a surprise. However, given the huge influence of Facebook and the mounting concerns about potential misuse of its information and platform, a significant risk is that Facebook will need to devote considerably higher resources to monitoring its users. If this occurs, then its COGS ratio would be expected to rise.

R&D expenditures have historically been large for Facebook, as might be expected for the advanced technology sector. These will remain a significant cost, but their future level is uncertain. Long-term annual revenue growth is also a key, and uncertain, driver of

	A	B	C	D	E	F	G	H	I	J	K	L	M	N	O	P
1	FACEBOOK INC A (FB) CashFlowFlag INCOME STATEMENT						Forecasts									
2	Fiscal year ends in December. USD in millions.	2013-12	2014-12	2015-12	2016-12	2017-12	2018	2019	2020	2021	2022	2023	2024	2025	2026	2027
3	Revenue	7872	12466	17928	27638	40653	48784	58540	70248	84298	101158	117950	133519	146604	155987	160666
4	Cost of revenue	1875	2153	2867	3789	5454	8216	9859	11831	14197	17036	19864	22486	24690	26270	27058
5	Gross profit	5997	10313	15061	23849	35199										
6	Research and development	1415	2666	4816	5919	7754	10412	12494	14993	17992	21590	25174	28497	31289	33292	34291
7	Sales, General and administrative	1778	2653	4020	5503	7242	10149	12178	14614	17537	21044	24537	27776	30498	32450	33424
8	Total operating expenses	3193	5319	8836	11422	14996										
9	Operating income	2804	4994	6225	12427	20203										
10	Interest Expense	56	23	23	10	6										
11	Other income (expense)	6	-61	-8	101	397										
12	Income before taxes	2754	4910	6194	12518	20594	20007	24009	28811	34573	41487	48374	54760	60126	63974	65893
13	Provision for income taxes	1254	1970	2506	2301	4660	6687	8025	9630	11556	13867	16169	18303	20097	21383	22025
14	Net income from continuing operations	1500	2940	3688	10217	15934										
15	Net income	1500	2940	3688	10217	15934	13320	15984	19181	23017	27620	32205	36456	40029	42591	43869
16	Key Ratios															
17	Revenue growth		0.5836	0.4382	0.5416	0.4709	0.2	0.2	0.2	0.2	0.2	0.166	0.132	0.098	0.064	0.03
18	COGS	0.2382	0.1727	0.1599	0.1371	0.1342	0.1684	averages								
19	R&D	0.1798	0.2139	0.2686	0.2142	0.1907	0.2134									
20	SG&A	0.2259	0.2128	0.2242	0.1991	0.1781	0.2080									
21	income tax rate	0.4553	0.4012	0.4046	0.1838	0.2263	0.3342									
22																
23	Teriminal Growth Rate	0.03														
24	Beta	1.14														
25	Riskless Rate	0.03														
26	Risk Premium	0.0596														
27	Cost of capital	0.097944														
28	PV net income (10 years)	$164,929														
29	Terminal Value	$645,659														
30	PV of terminal value	$253,630														
31	Total Value	$418,560														

FIGURE 3.3
Initial Facebook Valuation Results

the company valuation. Finally, the relative riskiness of investing in Facebook, rather than investing in the overall US stock market, is uncertain (and represented by beta in our model).

We will model the first three inputs using Pert distributions, which is a probability distribution that was developed for modeling expert opinion. An alternative that is sometimes used is the Triangular distribution. Figure 3.4 shows a comparison of both distributions for a revenue growth assumption with a minimum of 15%, most likely value of 20%, and maximum value of 30%.

Both the Pert and Triangular distributions use three parameters: the minimum, most likely, and maximum values. These three parameters are easy concepts to communicate to subject matter experts, and therefore are commonly used when soliciting expert opinion. However, experts generally have more difficulties identifying the minimum and maximum estimates compared with the most likely value. Part of the difficult is definitional—is the minimum an absolute minimum or a minimum "reasonable" value? As Figure 3.4 shows, with the same parameters used in both distributions, the Pert distribution is less sensitive to the minimum and maximum values than is the Triangular. This is the principal reason that we prefer the Pert to the Triangular: misjudgments regarding the minimum and maximum estimates will not affect the distribution (and the distribution's mean) as much.

However, rather than rely on one single expert's opinion, we have contacted three finance experts and asked them for the minimum, most likely, and maximum values for these three variables.[7] Each expert's opinion is then modeled using a Pert distribution. We have chosen to apply different weights to the three expert opinions, which in our example case are based on their expertise in the area of public company valuation. Figure 3.5 shows the section of the spreadsheet with this information.

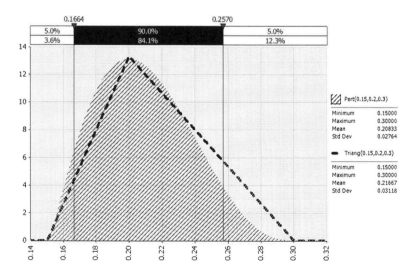

FIGURE 3.4
Comparison of the Pert and Triangular Distributions

	G	H	I	J	K	L	M	N	O	P
23	Expert Panel		min	ML	max	Pert	weight			
24	Rev Growt	#1	0.2588	0.3582	0.5086	0.3667	0.25			
25		#2	0.15	0.22	0.35	0.23	0.3	These 3 weights sum to 1.0		
26		#3	0.156	0.249	0.301	0.2421667	0.45			
27	COGS	#1	0.1300	0.1510	0.1684	0.1504	0.25			
28		#2	0.13	0.15	0.17	0.15	0.3	These 3 weights sum to 1.0		
29		#3	0.131	0.142	0.172	0.1451667	0.45			
30	R&D	#1	0.1798	0.1997	0.2134	0.1986585	0.25			
31		#2	0.2	0.25	0.3	0.25	0.3	These 3 weights sum to 1.0		
32		#3	0.18	0.2	0.233	0.2021667	0.45			

FIGURE 3.5
Expert Opinion Inputs to the Valuation Model

The final step for these three uncertain inputs is to combine the expert opinions. We do this through the use of the RiskDiscrete distribution: RiskDiscrete(L24:L26,M24:M26) for long-term revenue growth (with similar construction for COGS and R&D expenses).

Our probability distribution for beta is based on an estimate from the behavior of Facebook's stock price relative to the market average. Beta can be estimated by constructing a simple linear regression with the daily change in Facebook's stock price as a function of the daily change in a market index: the coefficient on the index provides an estimate of the relative variability of Facebook's stock to that of the overall market.[8] Values greater than 1 indicate greater variability (i.e., greater risk) for Facebook than for a market index. An inherent assumption of a simple linear regression is that the parameter uncertainty of the regression coefficients are normally distributed, so we use a Normal distribution with mean and standard error estimated from the regression model.

Running the simulation provides the valuation results shown in Figure 3.6, and the tornado chart in Figure 3.7.[9]

The mean value for Facebook is $668 billion, with a 90% confidence interval in the total future discounted FCFs of ($396 billion, $1266 billion). One may then wonder: Which of the four uncertain variables drives most of this uncertainty in future FCFs? The tornado chart answers this and reveals that revenue growth is the most influential source of uncertainty, followed distantly by R&D, beta, and finally COGS. The uncertainty in the long-term annual future revenue growth alone could drive the expected FCFs from $405M to $1299M. The uncertainty in the average future COGS only drives about $50M in FB's valuation (from $635M–$685M).

A few further notes regarding expert opinion are in order[10]:

- Provide experts with as much context and model design as possible in order to obtain more thoughtful and useful estimates.

- First, discuss scenarios in which a variable may be very low or very high, without discussing the actual quantitative estimate. This can help reduce "overconfidence," which is a very common bias.

- Related to this, ask the experts for reasons for their choices. This helps them think carefully about their estimates and may help with the choice of weights.

- Be careful when choosing experts. Select people with expertise, but with an open mind, and use techniques to reduce possible biases. Use more than one expert if possible, and weight their opinions appropriately. When using more than one expert, it is most useful if their estimates are obtained independently (use the "wisdom of crowds" but avoid "group think" and "anchoring," both of which are common biases, too).

- Show the expert what their input looks like and give them an opportunity to alter their judgment when they see the shape of the distribution.

- Consider using alternative parameters and/or probability distributions (there is also an "artist" tool in @Risk that permits construction of a distribution visually and interactively, although for most situations a Pert distribution will likely be sufficient).

- Ask experts to consider relationships (i.e., correlations) between the uncertain inputs (we consider this topic in Chapter 4).

Selecting Distributions

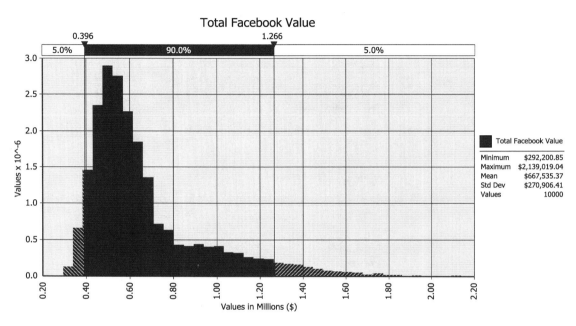

FIGURE 3.6
Facebook Simulation Results

3.3 Second Introductory Case: Value at Risk—Fitting Distributions to Data

The global financial crisis of 2007–2009 has focused attention on current practices of risk measurement and management in financial markets. Regulatory agencies throughout the world have adopted value at risk[11] and stress testing[12] techniques to measure the exposure faced by financial institutions. For example, the Basel Committee on Banking Supervision revisited the risk guidelines established by the Market Risk Amendment (updated in 2006). The current system adopted by the Basel Committee requires certain institutions to calculate a 99th percentile, left tail confidence interval in deriving the VaR measure. Management has some flexibility to choose among various methods for estimating this, including historic simulations and Monte Carlo simulations.[13] The chosen methods are back-tested against historical

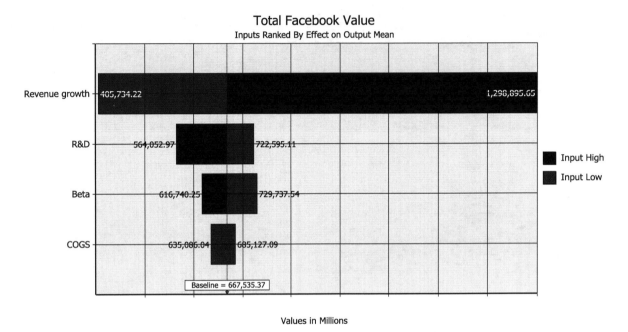

FIGURE 3.7
Tornado Chart for the Facebook Simulation Model

data to track the frequency with which actual losses exceed those that are expected based on the confidence interval. Capital adequacy is determined using, among others, the estimation of VaR.

Stress testing takes VaR one step further by asking about the potential size of losses that can occur outside of this confidence interval. So, in banking, the typical and basic task for both VaR and stress testing is to estimate the potential of losses of various sizes (at certain confidence levels) for a given portfolio of assets. Monte Carlo simulation is an excellent tool for establishing these estimates.

3.4 Distribution Fitting for Value at Risk—Parameter and Model Uncertainty

We explore estimating value at risk for a hypothetical portfolio of $1000 invested in a single stock.[14] Daily data is available in the spreadsheet VaR.xlsx for 3 years of daily stock prices.[15] Figure 3.8 shows the daily closing price and day-to-day change in the price of this stock. The daily change could be calculated as a percentage

Selecting Distributions 57

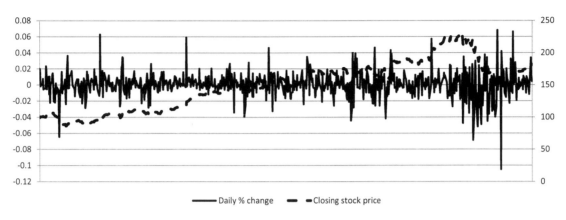

FIGURE 3.8
AAPL Daily Stock Price and Day-to-Day Change

return by dividing the difference in the daily closing prices by the beginning price (and multiplying by 100). Instead, we use the continuous version, the natural logarithm of the ratio of successive daily closing prices—often called the *log return*. These two calculations provide nearly identical results based on the fact that $\ln(1+x) \approx x$ for small values of x.

Time series patterns, such as those shown in Figure 3.8, are not uncommon for equities: the prices follow a meandering pattern, while the period-to-period changes fluctuate around the long-term trend.[16] Our focus will be on the VaR calculated over a 5-day period (these 5-day changes are calculated in Column D of the spreadsheet). From the historic data, the 5th percentile of the distribution is −4.98% (meaning that worst 5% of 5-day price changes was a decrease of 4.98% or greater) and the 1st percentile is −9%.

We will estimate the 5-day VaR for this stock holding. One approach is to use these historic returns to estimate the 95% and 99% VaR (the 95% VaR means that there is a 95% chance that the losses in a portfolio will not exceed the estimated amount within a certain time frame, for example, a month, or a day in this case—this corresponds to the 5th percentile of the price change distribution in that specific period). The use of historically observed losses, however, relies entirely on what has been observed during the time period for which data has been collected.

An alternative VaR estimation technique involves fitting a distribution to the historical price changes, and then simulating what the losses might look like over a specific period of time (we use 5 days in this example). VaR3.xlsx contains a number of models (of increasing complexity) that accomplish this.

The simplest approach is to fit a distribution to the 5 years of data. On the @Risk toolbar, click the drop-down menu under "Distribution Fitting," and choose the first option, "Fit." Enter the range for the 5-day price changes (D6:D756) and click "Fit" (alternatively, you can choose only a subset of distributions to fit—this is useful when your data set is very large and you already know which exact distribution(s) you'd like @Risk to fit to the data). @Risk will not allow some of the distributions to be fit—don't worry, as these distributions cannot be fit to this particular data (for example, the presence of negative values precludes some distributions, such as the Lognormal distributions, which can only be positive). @Risk will fit each appropriate distribution to the data[17] and after that compute several values called "goodness of fit" (GOF) statistics to describe how well each type of distribution fits the data. There are a variety of different GOF statistics, the first (and best) of which is the information criteria (IC): the Bayesian information criteria (BIC) and the Akaiki information criteria (AIC). Both of these ICs are based on estimating the "joint likelihood" that the observed data could have come from each postulated distribution, and, in addition, applying a penalty function for distributions with more parameters. The other GOF statistics are the chi-squared, the Anderson-Darling, or Kolmogorov-Smirnov statistics, which are generally less preferred (in addition, the chi-squared is only appropriate for fitting discrete distributions). @Risk lists the distributions from best fit to worst (you can choose which GOF statistic to use to rank the distributions, as they might rank differently for each criterion). Figure 3.9 shows the completed Fit Distribution window.

The Laplace distribution fits best (according to the BIC criterion), although several distributions are close—Figure 3.9 shows the data (as a histogram) together with Laplace distribution fit and Logistic distribution fit. You can click on each distribution in the list to visually see how it fits the actual data, and we recommend that you should certainly do so for a few of the best-fitting distributions. Regardless of the values of the IC, you should always visually check to ensure that the distribution appears to be a reasonable description of the actual data. You should also compare some of the statistics from the data with those of the distributions, for example, the mean, the standard deviation, and possibly some percentiles. Figure 3.10 shows a variety of summary statistics from the data and the two best-fitting distributions, along with the deciles of these distributions.

What you're looking for is that generally the distribution statistics are similar to those of the fitted distribution. After this, clicking the Write to Cell button will provide several options for placing the chosen distribution into the spreadsheet for simulation. We have placed the two best-fitting distributions, as well as the Ext Value Min distribution (which was the fifth best fit on the list), into the spreadsheet, as shown in Figure 3.11.

Note that we have also placed RiskPercentile functions next to each distribution—since we are interested in the first percentile as a result, we directly enter these into the spreadsheet. We will also mark each distribution as an Output so we can graphically examine the results. The results are shown in Figure 3.12.

Figure 3.12 uses the Summary Trend graph, changed to a box plot in which we have changed the settings to show the 1%, 99% values as the whiskers in the plot. The actual values read from the spreadsheet show the VaR measure (worst 1% of the simulation runs) as −8.2%, −6.9%, and −13.1% for the Laplace, Logistic, and Ext Value Min distributions, respectively.[18] Figure 3.12 shows several features of note. The Laplace distribution is less variable in the middle of the distribution (the 50% box) but a bit more variable in the tails of the distribution (as seen by the lower value for the first percentile) compared with the Logistic distribution. The Ext Value Min is much more variable. The historic first percentile (−9%) is lower than the first two distributions, but not nearly as bad as for the Ext Value Min distribution.

As you can see, the choice of distribution clearly matters for the results, which was also already visible when we fitted the distributions to the data and compared the various statistics.

Selecting Distributions

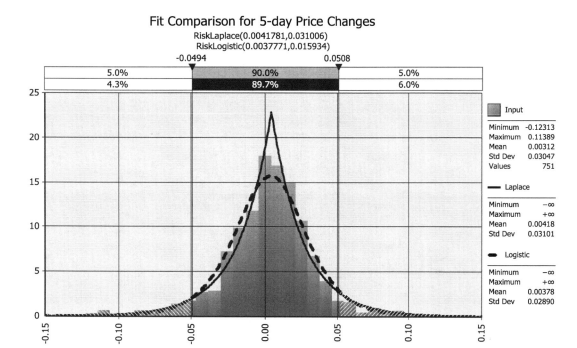

FIGURE 3.9
Distributions Fit for Daily Changes in AAPL Stock Price

3.4.1 Parameter Uncertainty (More Advanced, Optional)

We could complete a VaR analysis at this point, using the Laplace distribution, but let us compare it to a couple of additional modeling options. The best-fitting Laplace distribution has a particular mathematical form, and @Risk has helped us estimate the parameters of this distribution. In our example, the parameters are that mu = 0.0041781 and sigma = 0.031006. However, even though we

obtained the distribution with the parameters that best fit the data, we will have uncertainty about the *true* values of these two parameters, mu and sigma. If we think of the observed data as a random sample from an underlying process (which follows a Laplace distribution), then, assuming the underlying process doesn't change, if we observe another 3 years of stock price changes, we will get somewhat different data and thus different parameter estimates for mu and sigma. In other words, in both cases, the sets of random samples came from

	Minimum	Maximum	Mean	Mode	Median	Std Dev	Skewness	Kurtosis	10%	20%	30%	40%	50%	60%	70%	80%	90%
Input	-0.12313	0.11389	0.00312	≈-0.00461	0.00418	0.03047	-0.3103	5.3351	-0.02852	-0.01727	-0.0079	-0.00143	0.00418	0.00994	0.01611	0.02334	0.03479
Laplace	-∞	∞	0.00418	0.00418	0.00418	0.03101	0	6	-0.03111	-0.01591	-0.00702	-0.00071	0.00418	0.00907	0.01538	0.02427	0.03946
Logistic	-∞	∞	0.00378	0.00378	0.00378	0.0289	0	4.2	-0.03123	-0.01831	-0.00972	-0.00268	0.00378	0.01024	0.01728	0.02587	0.03879

FIGURE 3.10
Statistical Comparison of Distribution Fits

	A	B	C	D	E	F	G	H
1	Date -AAPL	Adj Close	Daily change	5 day change	5th percentile	=PERCENTILE.EXC(D6:D756,0.05)		
2	42443	97.351677			1st percentile	=PERCENTILE.EXC(D6:D756,0.01)		
3	42444	99.307831	=LN(B3/B2)				outputs	1st percentile
4	42445	100.627762	=LN(B4/B3)		Best Fit	=RiskLaplace(0.0041781,0.031006)	=RiskOutput("Best Fit")+F4	=RiskTheoPercentile(F4,0.01)
5	42446	100.466324	=LN(B5/B4)		2nd Best Fit	=RiskLogistic(0.0037771,0.015934)	=RiskOutput("2nd Best Fit")+F5	=RiskTheoPercentile(F5,0.01)
6	42447	100.580276	=LN(B6/B5)	=LN(B6/B2)	Ext Value Min	=RiskExtvalueMin(0.018134,0.032378)	=RiskOutput("Ext Value Min Fit")+F6	=RiskTheoPercentile(F6,0.01)
7	42450	100.570778	=LN(B7/B6)	=LN(B7/B3)				
8	42451	101.339943	=LN(B8/B7)	=LN(B8/B4)	2nd order			=AVERAGE(X2:X1001)
9	42452	100.779694	=LN(B9/B8)	=LN(B9/B5)			10th %	=PERCENTILE(X2:X1001,0.1)
10	42453	100.342873	=LN(B10/B9)	=LN(B10/B6)			90th %	=PERCENTILE(X2:X1001,0.9)

FIGURE 3.11
Distributions Ready for Simulation

the same underlying process, but the fitted distribution for each set will produce different parameters for the Laplace distribution. This uncertainty in the parameters (also called parameter uncertainty, or second-order uncertainty) can be estimated using various methods, including with the (parametric) bootstrap capability within @Risk. This is found under the Bootstrap tab in the Distribution Fitting window, as shown in Figure 3.13.

After @Risk has fit the distributions, click on the Bootstrap Analysis button at the bottom of the Fit Results window (second button from the right) and go to the parameter confidence intervals. Figure 3.14 shows this step.

To see the statistics on the right side of Figure 3.14, use the drop-down arrow on the right and choose the Data Grid. This shows 1000 bootstrap samples for the first parameter (mu) of the fitted Laplace distribution. The bootstrap samples are generated by simulating a new set of data points using the best-fitting Laplace distribution, and then refitting a Laplace distribution to these new data points (the bootstrap sample) and doing this 1000 times. Figure 3.14 shows the resulting estimates for mu (for 1000 such bootstrap samples). We refer to this as second-order uncertainty—we have retained the best-fitting distribution for the data (the Laplace distribution), but we have simulated that our empirical data points are just one

Selecting Distributions

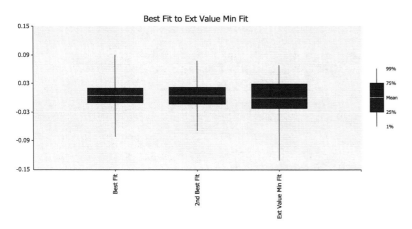

FIGURE 3.12
Simulation Results for 5-Day AAPL Price Change

possible sample, and thus we should estimate and consider our uncertainty about the true values of mu and sigma. The second-order uncertainty provides an estimate of how uncertain we are about the parameters' true values due to the fact that we only observe limited amounts of empirical data points. In other words, the more data that we have observed, the less the second-order uncertainty will be.

Choosing the second parameter (sigma) allows us to see the 1000 bootstrap estimates for the second parameter in the Laplace distribution. We copy and paste both sets of bootstrap samples into the spreadsheet, as shown (just the first 25 of the 1000 are shown in the figure) in Figure 3.15.

Columns U and V now contain the bootstrap samples of possible values of mu and sigma. Column W contains Laplace distributions with their parameters in Columns U and V: in other words, Column W contains

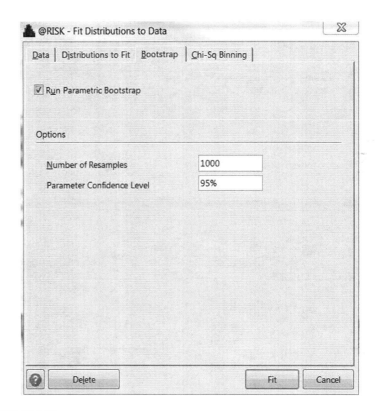

FIGURE 3.13
Bootstrap Option for Fitting Distributions

1000 Laplace distributions that all could have results in generating the original (empirical) data that we observed. We are interested in estimating the uncertainty around the first percentile of our best-fitting Laplace distribution, which we can do by estimating what the first

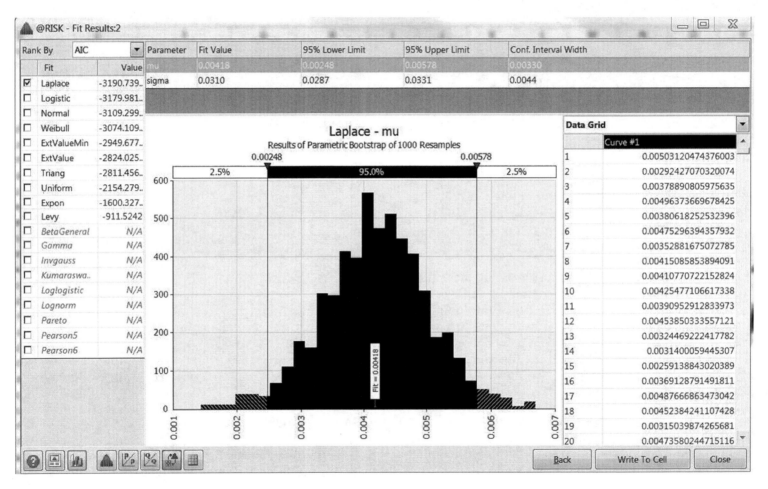

FIGURE 3.14
Bootstrap Parameter Confidence Intervals

	U	V	W	X
1	mu	sigma	bootstrap	1st percentile
2	0.005031	0.032069	0.005031	-0.0836783
3	0.002924	0.029765	0.002924	-0.0794112
4	3.79E-03	0.029328	0.003789	-0.0773382
5	0.004964	0.031089	0.004964	-0.0810359
6	0.003806	0.030997	0.003806	-0.0819389
7	0.004753	0.030935	0.004753	-0.0808196
8	0.003529	0.031536	0.003529	-0.0837062
9	0.004151	0.030739	0.004151	-0.0808812
10	4.11E-03	0.029031	0.004108	-0.0761975
11	0.004255	0.031411	0.004255	-0.0826355
12	0.00391	0.030125	0.00391	-0.079424
13	0.004539	0.03093	0.004539	-0.0810209
14	0.003245	0.03254	0.003245	-0.0867667
15	0.00314	0.031034	0.00314	-0.0827062
16	0.002591	0.031891	0.002591	-0.0856264
17	0.003691	0.031324	0.003691	-0.0829572
18	0.004877	0.029927	0.004877	-0.0779083
19	0.004524	0.031771	0.004524	-0.0833619
20	0.00315	0.03025	0.00315	-0.0805265
21	0.004736	0.032569	0.004736	-0.0853568
22	0.003983	0.031914	0.003983	-0.0842998
23	0.004231	0.032156	0.004231	-0.0847205
24	0.001765	0.030792	0.001765	-0.0834125
25	0.004379	0.032887	0.004379	-0.0865951
26	0.004284	0.030528	0.004284	-0.0801624

FIGURE 3.15
Bootstrap Samples

percentiles are of these possible Laplace distributions. In order to estimate the first percentile of each of the 1000 distributions, there is no need to run a simulation—instead, we use the RiskTheoPercentile function (this calculates the first percentile without needing to run a simulation), as shown in Column X. Figure 3.16 then shows a histogram for the 1000 first percentiles of the bootstrap Laplace distributions.

As seen from Figure 3.16, as we would have expected, the mean bootstrap 1 percentile estimate (i.e., the 1% 5-day VaR), −0.081509, is quite close to the 1 percentile from the best-fitting Laplace distribution, −0.08159. But the uncertainty in this 1-percentile estimate is what second-order uncertainty is about. The 90th percentile interval for this estimate ranges from −0.08673 to −0.07633. The purpose of the 1% 5-day VaR is to quantify the potential 5-day downside risks, and using the parametric bootstrap approach shown above, we were able to estimate how certain we are about the value of −0.08159. Especially when we have limited amount of data, using this approach can provide a more robust picture of risk than the single-point estimate from the best-fitting static distribution.[19]

Each of the curves in the bootstrap sample is a Laplace distribution, but with the uncertainty around each of the parameters included. This is sometimes called a *second-order distribution fit*, to be contrasted with *first-order distribution fitting*, which uses the single best-fitting set of parameters.

Typically, VaR estimates are based on the idea that the VaR is a single number, and not a range of possible numbers. It is not clear if the mean of the uncertainty distribution of 1% VaR values is the appropriate benchmark for measuring value at risk. Using the first percentile of the 1% VaR is probably too restrictive (for judging capital adequacy). But using the mean of the uncertain 1% VaR may not adequately protect against risk. This is an unresolved issue with the use of VaR.[20]

We used the VaR example to illustrate how parameter uncertainty can be included in a simulation model. While the "average VaR" was not much affected by including the additional uncertainty, our finding is that if there is limited relevant data available to estimate the VaR, the VaR also has uncertainty and can therefore be

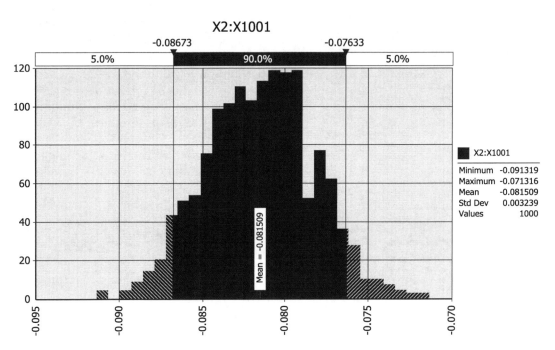

FIGURE 3.16
VaR (1%) for the Bootstrap Samples

represented as an uncertainty distribution. Taking into account such parameter uncertainty will typically result in more conservative VaR estimates.

VaR is typically expressed as a dollar value of how much of the investment (or portfolio) is at risk of loss. Translating our results into these units provides an estimate of VaR of $82 for a $1000 investment (in AAPL stock over a 5-day period). Our second-order uncertainty estimate ranges from $76 to $87.

While @Risk includes powerful techniques for fitting distributions to data and the ability to include and quantify the parameter uncertainty of the fitted distributions, we end with several important notes. First, is the data that you are using relevant to the problem that you are trying to model? When fitting a distribution to data, an assumption is that you believe that the "system" will behave similarly to the way it has in the past (i.e., no structural changes have happened). Is this a valid assumption? Second, it is important to always look at the distributions you are fitting and not rely solely on the information criteria measures for the "goodness of fit." You should also compare different statistics of your empirical data versus those same statistics of the fitted distribution (ideally, they are very close). You may be especially interested in a particular part of the distribution (such as the left-hand tail of the distribution, i.e., the lower percentiles), and so it may be the distribution that fits this part of the data best that is most relevant—but that may not be the distribution that fits best overall. Third, you should make sure the distribution you are fitting makes sense for the problem you are modeling. Each distribution possesses properties (such as the extent to which tail probabilities can occur) that may not be appropriate for your problem. *Quantitative measures of "the goodness of fit" cannot substitute for the use of your judgment on the appropriateness of the fit.*

3.4.2 Model Uncertainty (Most Advanced, Optional)

We could go one step further and explore sensitivity of VaR estimates to alternative models of uncertainty, that is, fitting several different probability distributions to the original (and to bootstrap samples) data. An example of this was our comparison of the Laplace, Logistic, and Ext Value Min distributions.

Considering the fact that there may also be uncertainty around which probability distribution is the correct/best distribution is part of a field called "model uncertainty." One approach of dealing with this model uncertainty is to build and run the model several times, each time with a different distribution, as an example of how much the results will change due to the choice of a different distribution. A key question is this case is: Would your decision change (and if so, how) if you used a different probability distribution? Another alternative approach is called "Bayesian model averaging," but this is beyond the scope of this book.

3.5 Third Introductory Case: Failure Distributions

Many phenomena are known (or believed) to follow particular probability distributions. For example, electronic components are often modeled using an exponential distribution—the use of this distribution assumes that the probability of failure during any time interval is constant (and that the occurrence of a failure is independent of past historical failures). The exponential distribution is a particularly simple one in that it has only one parameter, often referred to as the mean time between failures (MTBF).[21] Figure 3.17 shows an exponential distribution with a MTBF equal to 1000 hours.

The probability density shows the probability of failure as a function of time. Note that the probability is highest at the initial time, and monotonically declines with time. Since the exponential distribution assumes a constant conditional probability of failure (i.e., if failure has not already happened, the probability it happens during the next time interval is the same as in the last one), the highest probability occurs at the outset (probability = p), and at time $t = T > 0$, the probability at time T is given by $T - 1$ non-failures $(1 - p)^{\wedge}(T - 1)$ and the constant conditional probability of failure at time T, p. That is $p * (1 - p)^{\wedge}(T - 1)$. As T grows, this probability will therefore decline.[22]

We have data for 35 airplane cockpit systems with their associated MTBF rates.[23] Figure 3.18 shows this data (which is in Cockpit Failures3.xlsx), along with Column C, created using the formula = RiskExpon(MTBF).

In this chapter, we will model these component failures as independent probability distributions—in reality, there may be correlations (which we examine in Chapter 4). The correlations can be very important, since they can substantially increase the

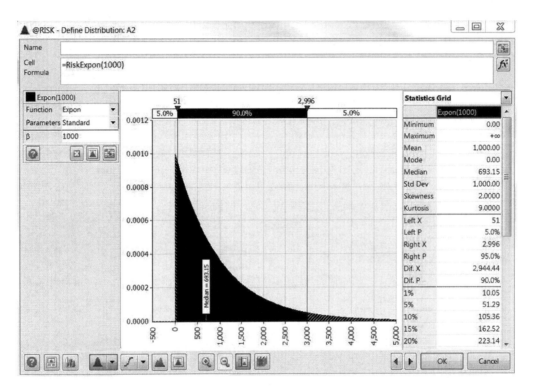

FIGURE 3.17
Example of an Exponential Distribution

time before one or more systems fail. When all 35 components are independent, we will refer to a system failure as when the first of the 35 components fails. With independent components, this turns out to be relatively quickly, despite the relatively long MTBF rates for each component individually. The shortest MTBF is 900 hours (four different components), but simulations reveal that at least one component will usually fail long before 900 hours. Figure 3.19 shows our model.

Each component's time to failure is modeled using an Exponential distribution, and the RiskTarget functions provide the probability

Selecting Distributions

	A	B	C
1	Event (system)	MTBF (hours)	Time to failure distribution
2	ADF Antenna	4000	4000
3	ADF Display	19900	19900
4	ADF Receiver	4200	4200
5	Airspeed Indicator 1	18100	18100
6	Airspeed Indicator 2	2500	2500
7	Alternator	7600	7600
8	Altimeter	5500	5500
9	Ammeter Vacuum Pressure	21500	21500
10	Attitude Indicator	2500	2500
11	Clock	17600	17600
12	Comm Antenna	1200	1200
13	Comm Radio	900	900
14	Directional Gyro	3400	3400
15	Fuel Quantity Indicator	16500	16500
16	GS Antenna	900	900
17	ILS Display	10000	10000
18	Left Fuel Quantity Transduce	51400	51400
19	LOC Antenna	900	900
20	Magnetic Comp	19900	19900
21	Marker Beacon Antenna	14800	14800
22	Marker Beacon Receiver	5300	5300
23	Oil Temp Pressure Gauge	6200	6200
24	Pilot System	73600	73600
25	Receiver	10000	10000
26	Right Fuel Quantity Transduce	51400	51400
27	Suction Gauge	21500	21500
28	Tachometer	8400	8400
29	Transponder	1700	1700
30	Transponder Antenna	9500	9500
31	Turn Coordinator	2100	2100
32	Vacuum Pump	4000	4000
33	Vertical Speed Indicator	145000	145000
34	VOR Antenna	9600	9600
35	VOR Display	10000	10000
36	VOR Receiver	900	900

FIGURE 3.18
Cockpit Failure MTBF Rates

that failure of at least one component (the MIN function is cell F1) occurs within the number of hours specified in cells E3:E26. We refer to this as the maintenance interval, since we would want to do maintenance work before any component fails. Running 10,000 simulations produces the results shown in Figure 3.20.

The mean failure time is 101.2 hours, and the results in the maintenance interval table show that there is more than a 20% probability of failure within a maintenance interval of 24 hours. These short times before system failure contrast starkly with the relatively long MTBF rates for each individual component. Again, among other reasons, this is dependent on the fact that we have modeled the individual component failures as independent. We will reexamine this in Chapter 5 and see that dependencies can greatly increase the times before system failure. In addition, in the current system, we assume there are no redundancies of any of the components, which means that the failure of *any one* of the 35 components would immediately result in failure of the full system.

3.6 Commonly Used Discrete Distributions

There are dozens of discrete and continuous probability distributions in @Risk. We will not provide an exhaustive description of each—documentation, including mathematic details and example models, is contained within the @Risk help file.[24] Many of these distributions are used on theoretical grounds—indeed, in many fields, the choices are made by understanding the system dynamics rather than by fitting to empirical data. Other distributions have very specialized applications. Our intent in this section (for discrete distributions) and the next

	A	B	C	D	E	F
1	Event (system)	MTBF (hours)	Time to failure distribution		first failure	=RiskOutput()+MIN(C2:C36)
2	ADF Antenna	4000	=RiskExpon(B2)		maintenance interval	failure probability
3	ADF Display	19900	=RiskExpon(B3)		1	=RiskTarget($F$1,E3)
4	ADF Receiver	4200	=RiskExpon(B4)		2	=RiskTarget($F$1,E4)
5	Airspeed Indicator 1	18100	=RiskExpon(B5)		3	=RiskTarget($F$1,E5)
6	Airspeed Indicator 2	2500	=RiskExpon(B6)		4	=RiskTarget($F$1,E6)
7	Alternator	7600	=RiskExpon(B7)		5	=RiskTarget($F$1,E7)
8	Altimeter	5500	=RiskExpon(B8)		6	=RiskTarget($F$1,E8)
9	Ammeter Vacuum Pressure	21500	=RiskExpon(B9)		7	=RiskTarget($F$1,E9)
10	Attitude Indicator	2500	=RiskExpon(B10)		8	=RiskTarget($F$1,E10)
11	Clock	17600	=RiskExpon(B11)		9	=RiskTarget($F$1,E11)
12	Comm Antenna	1200	=RiskExpon(B12)		10	=RiskTarget($F$1,E12)
13	Comm Radio	900	=RiskExpon(B13)		11	=RiskTarget($F$1,E13)
14	Directional Gyro	3400	=RiskExpon(B14)		12	=RiskTarget($F$1,E14)
15	Fuel Quantity Indicator	16500	=RiskExpon(B15)		13	=RiskTarget($F$1,E15)
16	GS Antenna	900	=RiskExpon(B16)		14	=RiskTarget($F$1,E16)
17	ILS Display	10000	=RiskExpon(B17)		15	=RiskTarget($F$1,E17)
18	Left Fuel Quantity Transduce	51400	=RiskExpon(B18)		16	=RiskTarget($F$1,E18)
19	LOC Antenna	900	=RiskExpon(B19)		17	=RiskTarget($F$1,E19)
20	Magnetic Comp	19900	=RiskExpon(B20)		18	=RiskTarget($F$1,E20)
21	Marker Beacon Antenna	14800	=RiskExpon(B21)		19	=RiskTarget($F$1,E21)
22	Marker Beacon Receiver	5300	=RiskExpon(B22)		20	=RiskTarget($F$1,E22)
23	Oil Temp Pressure Gauge	6200	=RiskExpon(B23)		21	=RiskTarget($F$1,E23)
24	Pilot System	73600	=RiskExpon(B24)		22	=RiskTarget($F$1,E24)
25	Receiver	10000	=RiskExpon(B25)		23	=RiskTarget($F$1,E25)
26	Right Fuel Quantity Transduce	51400	=RiskExpon(B26)		24	=RiskTarget($F$1,E26)
27	Suction Gauge	21500	=RiskExpon(B27)			
28	Tachometer	8400	=RiskExpon(B28)			

FIGURE 3.19
Cockpit Failure Model

section (for continuous distributions) is to review the most commonly used distributions and circumstances under which they are potential candidates for representing uncertainty about parameters in a model.

A number of discrete distributions are related to the *Bernoulli distribution*. A Bernoulli process is basically a discrete trial that has two possible outcomes, a "success" and a "failure." The probability of success, "p," is the only parameter of the Bernoulli distribution. The easiest example to imagine is tossing a coin. But any event where there are two possibilities can be modeled as a Bernoulli distribution—whether the government will enact a particular regulation, whether

Selecting Distributions

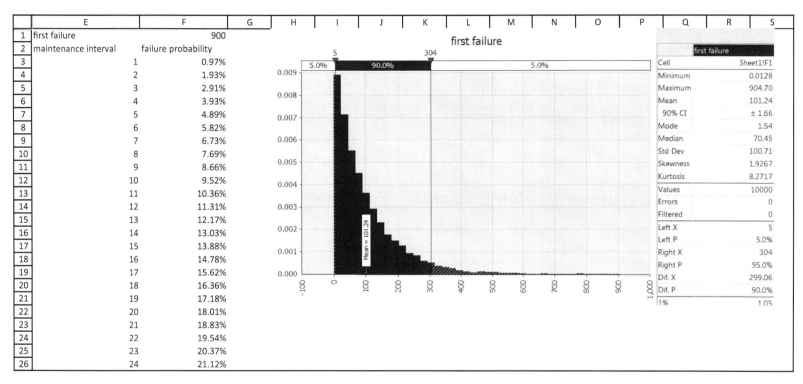

FIGURE 3.20
Cockpit Failure Probabilities as a Function of Hours

candidate A or candidate B will get elected, whether a drug gets approved by the FDA, whether the inflation rate at the end of the year will be greater than 3%, and so on.[25]

When a Bernoulli process is repeated a number of times, or *trials* (n), and each of the trials is independent and has the same probability of the event occurring (p), the resulting distribution for the number of successes out of n trials follows a *Binomial* distribution. For example, if you needed to model the number of people who drive red cars by calling 100 random phone numbers, the Binomial would be an appropriate choice.

A related distribution is the *Negative Binomial* distribution,[26] which gives the number of failures that occur before the sth success is obtained. A potential application would be to model the number of sales calls (each with a probability "p" of making a sale or not), that would be necessary before five sales (i.e., s = 5) are transacted.

The *Geometric* distribution provides the number of failures that occur before the first success is obtained, where each trial has the same probability of success ("p"). In other words, a Geometric distribution is a special case of the negative binomial distribution, but instead of looking for the sth success, it's looking for the first success. All of the distributions above are related to the Bernoulli and Binomial processes.

The *Poisson* distribution is conceptually related to the Binomial distribution. While the Binomial distribution assumes there are n trials, each with a probability of success p, within the Poisson process there is one parameter, lambda (λ), which is the constant *rate* at which successes occur. The Poisson is used when there is a continuing "exposure" that an event may happen, without the distinct (discrete) trials of the Binomial distribution. Example situations where a Poisson distribution is often used include modeling the number of accidents that happen within a certain period of time, or the number of outbreaks of a disease in a country over the next 5 years.

The *Discrete* distribution is quite different. It can be used to model a number of discrete values, each with some probability of occurring. For example, if future tax rates can take on one of five potential values (due to proposed legislation), then the discrete distribution can be used to simulate these outcomes, where the probability of each of the five possible scenarios needs to be specified by the user.

The relationship among several of these discrete distributions, and their relationship to the Bernoulli process, is illustrated in the spreadsheet Discrete3.xlsx. Imagine that we have 1000 manufactured items, where there is a constant defect probability (the spreadsheet uses p = 0.05). We can simulate each of the 1000 items, record whether it is defective, and calculate the total number of defects, the first defective item, and the number of items without defects before the first occurs.

1. The number of defects is the same as a Binomial (1000, 0.05) (you can check this by comparing the distribution of the number of defects to this Binomial).
2. The total number of defects can be approximated using a Poisson distribution (because n is large and p is relatively small).
3. The item with the first defect using a Negative Binomial (which is the same as a Geometric distribution because we're looking at the first defect). Figure 3.21 shows the portion of the spreadsheet (only the first 10 of the 100 items are shown in the figure) designed to do these calculations and simulations.

The simulation results shown in Figure 3.22 show that these distributions indeed closely resemble the actual calculations from the model.[27]

Figure 3.22 shows that the total failures and number of failures until the first defect can be simulated with many individual Bernoulli distributions, but they can also be estimated using the Geometric and Negative Binomial, and in this case, even the Exponential, distributions. While these distributions are all related, they will not always provide the same results, so care must be taken when choosing the appropriate one for modeling the problem at hand.

Selecting Distributions

	A	B	C	D	E	F	G	H	I	J	K	L
1	Prob	defective?	Item			Calculated						
2	0.05	=RiskBernoulli($A$2)	1				mean	std dev				
3		=RiskBernoulli($A$2)	2		Total #	=SUM(B2:B1001)	=RiskMean(F3)	=RiskStdDev(F3)				
4		=RiskBernoulli($A$2)	3		First	=VLOOKUP(1,B2:C21001,2,FALSE)	=RiskMean(F4)	=RiskStdDev(F4)				
5		=RiskBernoulli($A$2)	4		Number Failures	=F4-1	=RiskMean(F5)	=RiskStdDev(F5)				
6		=RiskBernoulli($A$2)	5		Failures out 30	=SUM(B2:B31)	=RiskMean(F6)	=RiskStdDev(F6)				
7		=RiskBernoulli($A$2)	6									
8		=RiskBernoulli($A$2)	7					Simulated				
9		=RiskBernoulli($A$2)	8				mean	std dev				
10		=RiskBernoulli($A$2)	9		Poisson	=RiskPoisson(A2*C1001)	=RiskMean(F10)	=RiskStdDev(F10)				
11		=RiskBernoulli($A$2)	10		NegBin	=RiskNegbin(1,A2)+1	=RiskMean(F11)	=RiskStdDev(F11)	Exponential	=RiskExpon(1/A2)	=RiskMean(J11)	=RiskStdDev(J11)
12		=RiskBernoulli($A$2)	11		Geometric	=RiskGeomet(A2)	=RiskMean(F12)	=RiskStdDev(F12)				
13		=RiskBernoulli($A$2)	12									

FIGURE 3.21
Model to Compare Distributions

3.7 Commonly Used Continuous Distributions

The *Normal* distribution (often referred to as a "bell-shaped curve" or *Gaussian* distribution) is the most widely known distribution. Quite a few naturally occurring phenomena follow a Normal distribution (but, careful, most do not!). The central limit theorem tells us that the parameter uncertainty around a sample mean for a random sample will be normally distributed (provided the sample size is large enough), and this is the basis for much of statistical inference. The Normal distribution has two parameters (the mean and standard deviation) and has the following characteristics: it is symmetric, with about 95% of the probability within +/− two standard deviations of the mean.

The *Lognormal* distribution is also characterized by a mean and standard deviation but has a right-hand skew. For example, the Lognormal may be a good choice for modeling the distribution of incomes within a population, or the price of a commodity 3 years from now.

The *Student-t* distribution is a similar distribution to the standard Normal (with mean = 0 and standard deviation = 1), but with fatter tails The greater the n (which is the parameter of a Student-t distribution and stands for the number of degrees of freedom), the more the Student-t distribution will look like a standard Normal.

The *Beta* distribution ranges from 0 to 1, has two parameters (alpha and beta), and can take a number of shapes depending on these parameters. It is commonly used to model the uncertainty about a proportions or prevalence (which is based on Bayesian statistics), such as the uncertainty about the probability of defective manufactured items, or the uncertainty about the prevalence of a disease.

The *Exponential* distribution has one parameter, β, which is the reciprocal of the mean of the distribution (i.e., $\beta = 1/\text{mean}$). It is

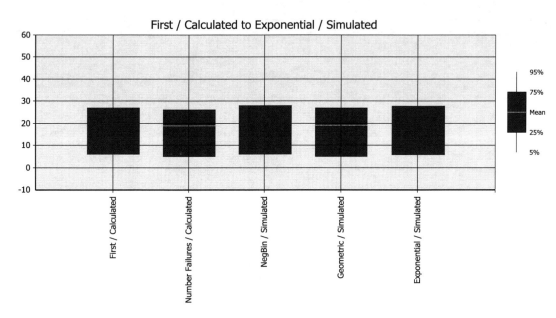

FIGURE 3.22
Distribution Comparison Results

also related to the Poisson: it gives the length of time until the first occurrence of an event, where the rate of the event occurring is constant over time (and equal to $1/\beta$). For example, the Exponential can be used to model the length of time, T, until the first call is made to a call center, while the Poisson could be used to model the probability that no calls are made during the time interval (0,T), or to forecast how many calls will be made per hour or per day.

The *Gamma* distribution is a generalization of the Exponential and provides the waiting time until the rth arrival, when the number of arrivals in each time period follows a Poisson distribution. The *Gamma* distribution has a minimum of 0, and its right-hand tail is typically skewed to the right. It is a flexible and commonly used distribution in a wide variety of applications and fields.

The *Pareto* distribution has two parameters and a similar shape to the Exponential (but with a heavier tail with higher probability of extreme outcomes). The Pareto distribution has the fattest tail of all distributions and is therefore often used in the insurance industry to model the potential impacts (e.g., costs) of extreme events.

Two distributions are commonly used to model the time until the next event, for example, to estimate the uncertainty of lifetimes of gadgets. The *Weibull* distribution also has two parameters (α and β) and provides the lifetime for an item with an instantaneous risk of failure that follows a power function (when $\beta > 1$, this means that the risk of failure will decrease over time). When $\beta = 1$, the Weibull is the same as the Exponential, and the instantaneous risk of failure is assumed to be constant. In other words, the Exponential distribution is a special case of the Weibull distribution.

The *Uniform* distribution, which has a minimum and maximum as its two parameters, may be used to model parameters in a model where there is no information to suggest whether some values are more likely than other values for a parameter. In other words, all values between the minimum and maximum are assumed to have the same value. One example of the use of the Uniform is how long you'll have to wait for the bus to arrive, if you have no information about the bus schedule. The Uniform distribution is also sometimes used to model expert opinion.

The *Triangle* distribution is commonly used to model expert opinion, and uses three parameters—the minimum, most likely, and maximum values. We generally prefer the use of the *Pert* distribution, however, because the mean and standard deviation of the Pert are less sensitive to extreme values for the minimum and maximum. For example, experts may provide extremely small minima or large maxima and, compared to the Pert distribution, the mean of the Triangle distribution is much more affected by these.

A number of distributions are available within @Risk for fitting distributions empirically to data without restricting them to any particular functional form (these distributions are called *non-parametric* or *empirical* distributions). Fitting these distributions is conceptually different than fitting "parametric" distributions because these distributions will exactly follow the data pattern, without imposing a preconceived functional form, such as a Lognormal distribution. Therefore, when fitting these distributions, there are no goodness of fit statistics to consider. Among these, commonly used are the *Cumulative* distribution (RiskCumul or RiskCumulD, depending on whether the data is ascending or descending in order), and the *General* distribution. The two cumulative distributions use as inputs a series of data values and the cumulative probability of observing values less than or equal to these (with the data values either in ascending or descending order). These distributions fit the data empirically, without imposing any assumptions regarding the underlying form of the distribution function.

3.8 A Brief Decision Guide for Selecting Distributions

A fundamental choice concerns whether distributions are to be chosen based upon the data or based upon judgment. Even if you have historical data, the choice must be faced if, for example, the data is relevant, accurate (e.g., without great biases), and accessible. Fitting past data, however perfect and however much uncertainty about the fit is included, can only model what the past data tells us. If there are reasons to expect the future to differ from the past, or the data may be biased, then some degree of judgment may be called for.[28] When judgment is used, it may be based on theoretical considerations (such as using an Exponential or Weibull distribution to model waiting times), as in Section 3.5 or 3.6, or expert opinion, as discussed in Section 3.2.

When fitting data, there are two primary approaches: parametric (functional) or empirical (non-parametric). Parametric fitting entails examining which probability distribution(s) best describes the uncertain process being modeled. When fitting data empirically, the task is to examine which probability distribution from a set of distribution functions best fits the data. The approach can be first order or second order to represent the uncertainty of the estimated parameters of the distribution that is fitted.

Non-parametric fitting (also referred to as empirical distributions) requires several additional steps. @Risk has three empirical distributions: Cumul, CumulD, and General. All require the data to first be sorted (ascending for the Cumul or General, and descending for the CumulD). A column of cumulative probabilities is needed—if you are using the full set of data values, each value can have the probability = 1/n, where n is the number of data points (if you have a lot of data, you can bin the data first, as you would to have Excel create a histogram). Since these functions require the cumulative probabilities, create a new column to sum these probabilities as you move through the sorted data. These empirical functions also require a minimum and maximum value—you can use the minimum and maximum values found in the actual data (or more extreme values, based on subjective judgment). Finally, these distributions are found under Define Distribution rather than Distribution Fitting. The reason is semantic: fitting refers to fitting a particular mathematical function to the data, while defining refers to letting the data define the distribution.

Figure 3.23 provides a schematic of the decision process for selecting distributions.

Regardless of the approach chosen, you should always visually examine the chosen distribution to ensure it either represents the data or judgment accurately. Sensitivity analysis for the chosen distributions and their parameters is also appropriate in order to understand the extent to which the choice of the distribution matters.[29] As former US Secretary of Defense Donald Rumsfeld stated, "There are known knowns. These are things that we know. There are known

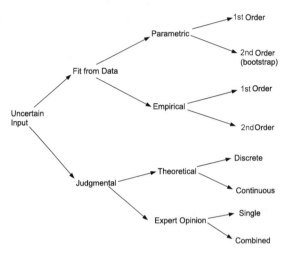

FIGURE 3.23
Decision Guide for Choosing Distributions

unknowns. That is to say, there are things that we now know we don't know. But there are also unknown unknowns. These are things we do not know we don't know."[30] We can model the known knowns and known unknowns, but modeling the last category is a lot harder.

Exercises

Exercise 3.1 Smoking and Birthweights

Exercise3.1.xlsx contains data on the birthweights (in ounces) of 484 babies born to mothers who smoked and 742 babies born to mothers who did not smoke.

a. Fit distributions to each set of baby birthweights.

b. Low-birthweight babies (defined to be less than 88 ounces and greater than or equal to 53 ounces) are at higher risk for a number of health problems. Very low-birthweight babies (defined to be less than 53 ounces) are generally premature and at significant risk to not survive. Estimate the risk of low and very low birthweight babies for smoking and nonsmoking mothers. The *odds* of an event are defined as the ratio of the probability of the event occurring to it not occurring. The *odds ratio* between two groups, then, is the ratio of the odds for the two groups. Express your answer in terms of the odds ratio of having a (very) low-birthweight baby for smokers versus nonsmokers.

c. Redo parts (a) and (b) using fitted normal distributions for both groups. Explain why the answers are different than before.

Exercise 3.2 Household Car Travel in the United States

The Federal Highway Administration conducts the National Household Travel Survey (NHTS) every 8 years.[31] Exercise3-2.xlsx contains the 2017 data for the sample of households in each state that own cars. For each sampled household in each state, the total annual miles driven is provided.

a. Fit distributions for each state—use the Batch Fit option found under Distribution Fitting. It will take a while, as @Risk will fit each distribution to the data for each state. When it is finished, there will be a summary worksheet showing the best-fitting distribution for each state.

b. Describe the most common best-fitting distributions across the states.

c. Produce a histogram showing the mean miles driven in each state. Do the same for the standard deviations.

d. Use the best-fitting distributions (according to the AIC) for CA, CO, NY, OR, OK, TX, and WI. Simulate these distributions and estimate the fraction of households in each of these states that drive less than 10,000 miles per year and the fraction that drive more than 20,000 miles per year.

Exercise 3.3 Investment Returns

Exercise3-3.xlsx provides the annual returns to stocks (measured by the S&P 500), Treasury bonds, and Treasury bills over the 1928–2018 period. Consider an investment of $100 invested in each of these assets in 2018.

a. Fit a distribution to each of these asset's annual returns. Choose distributions based on the best (lowest) AIC score.

b. Simulate the value of a hypothetical $100 investment in each asset by year over a 10-year period. What are your main assumptions in this model?

c. What is the probability of losing money in the investment in each asset over each time period?

d. What is the first percentile in each case?

e. Repeat a through d using a non-parametric distribution (RiskCumul).

f. Compare the results. Interpret your findings.

Exercise 3.4 Beer Sales

Exercise3-4.xlsx has 1 year of daily beer sales (in dollars) at a local microbrewery.

a. What revenue of daily beer sales is exceeded on 5% of days?
b. Fit a parametric distribution to daily sales and estimate the 95th percentile of that distribution. What assumption do you make when you fit a probability distribution to daily beer sales data? Do you think this is valid, and how could you examine this?
c. Fit a non-parametric distribution and determine the 95th percentile for daily beer sales.
d. For the best-fitting distribution (part [b]), estimate the second-order uncertainty for the 95th percentile of daily beer sales.
e. Compare the results for parts (a) through (d) with a table and/or graph.
f. The data in this exercise is real, but this part is hypothetical. Imagine that this microbrewery used a promotional effort on days 180–240. Assess the likelihood that this promotion had increased beer sales—provide an estimated probability that sales improved throughout this period, and the potential amount of such an increase. For this exercise, you should assume that there is no seasonality or trend in the daily beer-sales (we'll discuss this in Chapter 5). What additional assumptions are necessary for your answer to this question?

Exercise 3.5 Global Warming Predictions

Exercise3-5.xlsx provides several predictions regarding sea level rise (measured in meters) for the year 2100.[32] Global climate modeling is complex and there are many probabilistic scenarios developed, each of which accounts for different climate factors (CO_2 concentration, ice cap melting, permafrost conditions, etc.). The spreadsheet provides simplified results from a large number of studies, many of which have been conducted by the Intergovernmental Panel on Climate Change (IPCC).

a. Combine these surveys as expert opinions, using Pert distributions, each with equal weights. Some studies provide 90% confidence intervals, while others provide 66% confidence intervals, so use the Alternate Parameters to enter these distributions.
b. Combine these surveys, weighting more recent surveys more than older surveys (i.e., use weights proportional to their time currency).
c. A commonly used benchmark is a 2°C rise in the global temperature—which might result in a rise of 0.5 meters in the global sea level (although this is itself subject to considerable uncertainty [which we'll ignore in this exercise]). Using the two different weightings, estimate the confidence of a sea level rise of this magnitude or greater for both weighting schemes. What assumptions are you making in either of the two analyses?

Exercise 3.6 Electricity Prices

Exercise3-6a contains the average electricity prices for 2017 across the United States and by retail sector (data is from the Energy Information Administration[33]), in cents per kilowatt hour.

a. Fit distributions to the prices for each sector across the United States (use the Batch Fit option under Distribution Fitting). Compare the means of the fitted distributions with the averages for the entire United States (Row 54 of the spreadsheet).
b. Use the population data in the file Exercise3-6b to produce weighted average electricity prices for the United States

based on the best-fitting distributions, and compare this with the actual US average prices.

c. Are there any potential uses for the fitted distributions rather than the actual US averages?

Exercise 3.7 Warranties

Tires can fail for a number of reasons, such as tread wear-out, sidewall failure, or steel shear of tread. Assume that sidewall failures and steel shear of tread follow Exponential distributions, with the means equal to a design mileage chosen by the manufacturer. Tread wear-out follows a Normal distribution, with the same design mean, and a standard deviation of 5000 miles. The tire warranty mileage is set at 10,000 miles less than the design mileage. The warranty will cover all three types of damages, but it does not cover punctures. Punctures can be assumed to follow an exponential distribution (with mean time to failure of 35,000 miles). In addition, you can assume that the warranty is void if the car is sold to another owner, which can be assumed to follow an exponential distribution with a mean of 75,000 miles). Assume that a damaged tire must be replaced and consider only originally purchased tires (not tires replaced after a warranty covered failure—i.e., repeat failures).

a. The tire manufacturer can design tires for mean failures of 40,000, 60,000, or 110,000 miles. Estimate the probability of at least one warranty-covered failure for each design option (i.e., the probability of a covered failure occurring at a mileage less than the warranty mileage but before the car is sold).

b. Suppose the cost of producing the tires is $30 each for the 40,000-mile design, $50 for the 60,000-mile design, and $75 for the 110,000-mile design. A warranty replacement costs the manufacturer the cost of the new tire plus $30 of labor. The tires can be sold for $65, $95, and $120, respectively. Which design has the highest expected profit?

Exercise 3.8 Let's Make a Deal

a. Monty Hall was the host of the long-running television game show, *Let's Make a Deal*. The Monty Hall problem is a well-studied probability problem that continues to confound people with the results. At the end of each show, the day's winner has a final opportunity to win a great prize. They get to select one of three doors—only one of the doors has the great prize behind it, the other two have nothing valuable (often called "goats"). After the contestant chooses their door, Monty *always* reveals one of the two goats (never what is behind the door the contestant has chosen) and offers the contestant the opportunity to then change their choice. Should they change their choice?

Build a simulation model to illustrate the two possible strategies: Sticking with the initial choice or changing the initial choice. Estimate the probability of winning the great prize under each of these two strategies.

b. There have been many studies of the "hot hand" in sports (or stoc picking, for that matter). One paper claims to prove that there is a subtle and substantial bias in these studies.[34] The claim is that people misunderstand the nature of streaks, thereby erroneously testing for them.

As an example, consider flipping a coin 100 times. If you write down the outcome of the flip following a heads, finish the flips, and then see what fraction of the outcomes are heads, you might expect to find that proportion to be 50%. But it will not be! The expected proportion of heads following a heads will be less than one-half.

The simplest case to consider is three coin flips. There are only eight possibilities for what happens when you flip a coin three times. Each possible sequence is equally likely. But the expected proportion of heads that follow getting a heads will be 5/12 and not 1/2. Build a simulation model to show this.

Notes

1. A firm's riskiness in the CAPM model is a measure of "correlation," typically defined as the covariance between the firm's stock and the market divided by the variance of the market.
2. Similar assumptions were used in the Facebook valuation model provided by Aswath Damodaran (downloaded from http://aswathdamodaran.blogspot.com/2018/04/the-facebook-feeding-frenzy.html). While revenue growth has exceeded 50% per year, on average, it is nearly impossible to sustain that large of a growth rate as a company becomes as large as Facebook, so we have made a more modest assumption.
3. We use the cost of equity as the weighted average cost of capital because Facebook has very little debt.
4. Using CAPM, Cost of Equity = (Market Risk Premium*Beta) + Riskless Rate, i.e., (5.96%*1.14) + 3% = 9.79%.
5. It is worth noting that the valuation model certainly does not need to match the market valuation. One reason for building a valuation model is to see if the market may be under- or over-valuing a company. But if the initial model differs too much from the market valuation, it may cause the analyst to reexamine their base case assumptions.
6. In fact, the use of prediction markets is based on the notion that the "wisdom of crowds" can produce more accurate assessments of uncertain events than would result from any particular expert's opinion. For examples, see Spann, M. and Skiera, B. 2008. "Sports Forecasting: A Comparison of the Forecast Accuracy of Prediction Markets, Betting Odds and Tipsters." *Journal of Forecasting*, 28:1. Another example can be found in Kaufman-Scarborough, C. et al. 2010. "Improving the Crystal Ball: Harnessing Consumer Input to Create Retail Prediction Markets." *Journal of Research in Interactive Marketing*, 4:1. We illustrate the use of multiple opinions in this example.
7. Note that other parameters (alternate parameters, in @Risk terms) could be used, such as P10, P50, and P90 for the 10th, 50th, and 90th percentiles, respectively. The choice of parameters for soliciting expert opinion should not be overlooked—depending on the context and experts, some parameters may be more easily and reliably estimated than others.
8. We will postpone discussion of linear regression modeling until Chapter 4, but the regression model is shown on the FB tab of Valuation3b.xlsx. While this is a common way that beta is estimated, there are different approaches concerning whether to use daily, weekly, or monthly data as well as the appropriate time frame to use. Further, beta is a forward-looking risk concept, but is estimated from past data, so there are further adjustments that are made by various experts. If you look at the best estimates publicly available (for instance from https://finance.yahoo.com), it is not likely that you can derive that precise value from any permutation of time periods used to estimate it from stock prices.
9. The default in @Risk is to display all distribution inputs in the tornado chart. If there are many (or you only want to see some input distributions), then you can insert RiskCollect() into the distributions and under the Sampling tab of the Simulation Settings and change the dropdown menu for Collect Distribution Samples from "All" to "Inputs marked with Collect." We did that for Figure 3.7 in order to focus on the four major inputs and not on the individual expert opinion distributions.

10. The book *Superforecasting: The Art and Science of Prediction* by Dan Gardner and Philip E. Tetlock (Broadway Books, 2016) gives much more insight and many pieces of advice.
11. There are a number of definitions for value at risk. In general, VaR can be thought of as a measure of how low or high a certain metric may become (for example, how low a future loss from a portfolio may be), at a specific confidence level and within a specific period of time, for example, at the 5% level (i.e., once every 20 years) over a 1-year time period.
12. Stress testing also has a number of definitions in the context of a financial institution. In general, stress testing is the process of examining the behavior of a financial instrument, portfolio, or institution under extreme and crisis scenarios (liquidity, inflation, interest rate shocks, economic depression, etc.).
13. Basel Committee on Banking Supervision, "Minimum Capital Requirements for Market Risk," January 2019. It specifies that the VaR backtesting "may permit banks to use models based on either historical simulation, Monte Carlo simulation, or other appropriate analytic methods" (Section 33.9).
14. VaR is generally estimated for a portfolio of assets. Analyzing VaR for a portfolio follows the same methodology adopted here, but with the additional estimation of correlation among the assets, which we discuss in Chapter 4. We will also ignore correlation of price changes over time (autocorrelation), which we discuss in Chapter 5.
15. The stock is AAPL (Apple) and the time period is 1/29/2018–1/29/2019.
16. Of course, if there were any discernible trend in the changes that is predictable for the future, investors would have acted on these and the predictability of the trend would disappear. At least this is the assumption of the *efficient market hypothesis* (EMH). Even under the EMH, the volatility of the time series of changes may still exhibit predictable patterns, but you can ignore this in the present chapter. Chapter 5, dealing with time series analysis and forecasting, will permit us to model volatility patterns in the data.
17. @Risk uses maximum likelihood estimation (MLE) to estimate the parameter of all of the "fitted distributions."
18. We could have used the RiskPercentile function to derive these values after the simulation. Instead, we used RiskTheoPercentile functions to calculate the first percentiles of these distributions. In other words, it is not necessary to run a simulation to derive these values. Figure 3.12 shows the simulation results, while Figure 3.11 shows the RiskTheoPercentile functions we used for direct calculation.
19. As mentioned, estimating second-order uncertainty becomes more important the smaller the data set. With a large amount of data, the bootstrap results will only vary slightly from the static best fit (as in this case), though they still show uncertainty. But with small data sets, the uncertainty around parameters, and thus also the VaR, can be quite large.
20. An even more uncertain view of VaR can be found in Hsu, J. C. and Kalesnik, V. "Risk-Managing the Uncertainty in VaR Model Parameters." Chapter 18 in Gregoriou, G. 2009. *The VaR Implementation Handbook*. McGraw-Hill. These authors suggest the use of a variety of expert judgments in representing uncertainty about return distribution parameters. This would be an excellent application for use of the combined expert opinion (Section 3.2) in a VaR model. Instead of uncertainty about the parameters of the distribution being estimated from the historical data, they could vary according to expert opinion (similarly, the actual distribution itself could also be subject to differing expert opinions).
21. While the exponential distribution is sometimes used to model failure probabilities, the Weibull distribution is a more general and flexible (and usually more accurate) representation of actual failures (for example, see Wang, R., Sha, N., Gu, B. and Xu, X. 2014. "Statistical Analysis of a Weibull Extension with Bathtub-Shaped Failure Rate Function." *Advances in Statistics*, http://dx.doi.org/10.1155/2014/304724). Failure probabilities are often bathtub-shaped, where the probability falls during an initial break-in period and remains constant for an extended period of time after initial break-in, finally increasing later in the life cycle. In this example, we assume the probability to be constant and therefore can use an exponential distribution.

22. Electronic components often have a burn-in period for a related reason: the conditional probability of a failure declines over time during the early time periods of its life. A typical pattern is for this decline to be followed by a longer period of approximately constant probabilities of failure, followed by increasing probabilities in later periods. This "bathtub-shaped" failure probability pattern can be modeled using a Weibull distribution (the exponential is a special case of the Weibull and approximates the middle portion of this bathtub-shaped curve).
23. Lenz, M. and Rhodin, J. 2011. *Reliability Calculations for Complex Systems*. Department of Electrical Engineering, Linkopings Universitet, LiTH-ISY-EX-11/4441-SE. The data is given in Appendix B (attributed to a NASA report).
24. Other references provide detailed analysis and application of a number of these distributions, such as Bean, M. 2001. *Probability: The Science of Uncertainty*. Wadsworth Group.
25. This last example suggests an interesting place to obtain estimates of probabilities for Bernoulli distributions. Online prediction markets, such as PredictWise (www.predictwise.com), readily provide market-based estimates for probabilities of many financial, political, and social events. Several companies use internal prediction markets to estimate probabilities based on the collective judgment of their various stakeholders. There is even a journal, *The Journal of Prediction Markets* (published by the University of Buckingham Press), devoted to the study of such markets. Prediction markets also provide data for distributions other than the Bernoulli, depending on the event being marketed (e.g., there may be a series of markets for various ranges for future oil prices that taken together can approximate a continuous probability distribution). In addition, there is a long history of applications of the Delphi technique to elicit expert opinion from a group, although Delphi techniques generally attempt to build a consensus opinion out of a diverse group of opinions, instead of embracing the range of, and uncertainty in, opinions.
26. A negative binomial distribution is a special case of the Polya distribution. The latter is used frequently in insurance modeling.
27. Figure 3.22 was produced by marking the simulated cells as outputs and overlaying the distributions using the Summary Box Plot. This also illustrates how simulation may be used to derive probabilistic relationships. The relationships among these distributions can be derived mathematically, but they can also be verified empirically through simulation.
28. If there is sufficient data, a hybrid approach can be used where a distribution is fit to the data first, and then adjustments, based on judgment, are employed to modify the fitted distribution. We provide an example of this approach in Chapter 4.
29. In other words, if you are uncertain about the right distribution, you can always run the analysis multiple times, each time with a different distribution, to see how this may change the results and how it may change the best decision.
30. February 12, 2002, A Department of Defense news briefing, as reported by *Slate* in "The Poetry of D. H. Rumsfeld," by Hart Seely, April 2, 2003, accessed at https://slate.com/news-and-politics/2003/04/the-poetry-of-donald-rumsfeld.html on March 14, 2019.
31. https://nhts.ornl.gov/.
32. Data adapted from "Supporting Information For 'Evolution of 21st Century Sea-Level Rise Projections," Garner, A. J., Weiss, J. L., Parris, A., Kopp, R. E., Horton, R. M., Overpeck, J. T. and Horton, B. P. 2018. "Evolution of 21st Century Sea Level Rise Projections." *Earth's Future*, 6: 1603–1615. https://doi.org/10.1029/2018EF000991.
33. https://www.eia.gov/electricity/data/state/.
34. Miller, J. and Sanjurjo, A. 2018. "Surprised by the Hot Hand Fallacy? A Truth in the Law of Small Numbers." *Econometrica*, 86(6): 2019–2047.

4

Modeling Relationships

Learning Objectives

- Appreciate the importance of including and accurately reflecting key relationships in simulation models.
- Learn how logical functions can be used to capture important relationships.
- Learn how to include rank-order correlation structures between variables.
- Learn how to capture more complex correlations through the use of copulas.
- Learn how to fit correlation patterns to data.
- Learn how to use regression models to capture relationships in your models.
- See an example of how to model the uncertainties inherent in regression models.

4.1 First Example: Drug Development

Pharmaceutical companies invest considerable sums of money and time in the highly risky process of developing new drugs. Drug development goes through a number of stages, both for product development and regulatory approval, with each stage subject to many uncertainties, including whether the drug appears to be successful at treating a condition and whether there are adverse side effects that would prevent ultimate approval for use. These different development stages are clearly and logically related: a company will *only* undertake the next stage of development if previous stages were all successful.

These types of conditional and logical relationships are perhaps the simplest and most frequently encountered when building models. Most can be modeled using the variety of logical functions available within Excel: principally the IF, OR, AND functions and a number of related derivatives (such as VLOOKUP, COUNTIF, etc.). When

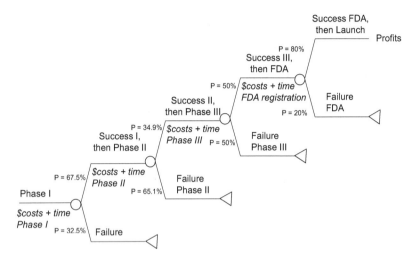

FIGURE 4.1
The Drug Development Process

considering our simple drug development example, these types of conditional relationships can be modeled using a combination of Bernoulli distributions and IF statements. Figure 4.1 shows this logic of the drug development process, formatted as a simple decision tree.

Figure 4.1 shows a typical drug going through three stages (phase I to III), followed by registration and potential approval by the FDA. This last step is, of course, conditional on it having survived all of the previous phases. The probabilities are from DiMasi et al. (2010) for the therapeutic class of gastrointestinal/metabolism drugs, based on self-originated compounds first tested on humans from 1993 to 2004.[1]

In addition to the probability of success of each development phase, each phase also has an associated (uncertain) cost and (uncertain) duration. The costs and durations are modeled using Pert distributions, and success (or failure) at each phase is modeled using a Bernoulli distribution.[2] Potential profit for the drug is also modeled using a Pert distribution.[3] The model (Drug Development4.xlsx) is shown in Figure 4.2.

While each phase's success or failure is simulated using the Bernoulli distribution, the formulas in Row 27 ensure that there is only a cash flow if the current phase and all previous phases are successful. If any phase's progress produces a 0 (i.e., the phase was not successful), all subsequent stages will show that there are no costs, revenues, and cash flow. Similarly, the profit calculation (Column G) contains an IF statement that requires C26*D26*E26*F26=1: all phases and registration must be successful in order to launch the drug and thus generate a profit.

The IF statements capture the relationships between the uncertainties about whether the drug will progress through each stage of the development process. Alternatively, we could also have used a combination of nested functions within a single cell: IF(AND(D26=1,E26=1,F26=1,G26=1), to model the dependence of product launch and profit on the success at all previous development stages. Multiplying the outcomes (either 0 or 1) of the cells modeling each previous stage accomplishes the same thing, but is simpler and less prone to entry errors.

Our model contains 13 probability distributions (four uncertain costs, four durations, four success or failure events, and the uncertain profit), and our main output is the NPV (including the expected NPV, often called eNPV) of this drug development.[4] Figure 4.3 provides the resulting simulation results for the NPV: the probability of losing money is over 90%.[5]

The expected NPV is only $−4.77 million, but the distribution is quite skewed (the simulated NPV ranges from −$72 million to +$181 million. Also evident in Figure 4.3 is that the NPV has multiple peaks—these

Modeling Relationships

FIGURE 4.2
Drug Development Model

	A	B	C	D	E	F	G	H
1		**Inputs**						
2								
3				Costs (in US$ millions)				
4		Phase I	Phase II	Phase III	Registration		Profit (in US$ millions)	
5	Distribution	=RiskPert(B6,B7,B8)	=RiskPert(C6,C7,C8)	=RiskPert(D6,D7,D8)	=RiskPert(E6,E7,E8)		Distribution	=RiskPert(H6,H7,H8)
6	Low	1	3	15	0.25		Low	75
7	ML	2.5	8	30	0.5		ML	200
8	High	10	15	100	1		High	500
9								
10				Duration (yrs)			Discount rate	0.12
11		Phase I	Phase II	Phase III	Registration			
12	Distribution	=RiskPert(B13,B14,B..)	=RiskPert(C13,C14,C15)	=RiskPert(D13,D14,D15)	=RiskPert(E13,E14,E15)			
13	Low	1	1	1.5	0.3			
14	ML	1.5	2	2.5	0.75			
15	High	2.5	3	4	1.5			
16								
17				Probability of succes (%)				
18		Phase I	Phase II	Phase III	Registration			
19	Probability	0.675	0.349	0.5	0.8			
20								
21		**Model**						
22								
23			T					
24					Model			
25		Phase	I	II	III	Registration	Profit	Total DCF
26		Progress?	=RiskBernoulli(B19)	=RiskBernoulli(C19)	=RiskBernoulli(D19)	=RiskBernoulli(E19)		
27		Cash flow	=-B5	=IF($C$26=1,-$C$5,0)	=IF($C$26*$D$26=1,-$D$5,0)	=IF($C$26*$D$26*$E$26=1,-$E$5,0)	=IF($C$26*$D$26*$E$26*$F$26=1,$H$5,0)	=IF($C$26*$D$26*$E$26*$F$26=1,$H$5,0)
28		Time	=0	=$C$28+$B$12	=$D$28+$C$12	=$E$28+$D$12	=$F$28+$E$12	
29		DCF	=$C$27/(1+$H$10)^$C$28	=$D$27/(1+$H$10)^$D$28	=$E$27/(1+$H$10)^$E$28	=$F$27/(1+$H$10)^$F$28	=$G$27/(1+$H$10)^$G$28	=RiskOutput("Total DCF")+SUM($C$29:$G$29)
30							eNPV	=RiskMean(H29)

correspond to the various times that the drug can fail during its research and development (R&D). The tornado chart in Figure 4.4 shows the sensitivity of the NPV to the various uncertain factors.

Most critical to the variability of the NPV are phase III and registration success/failure, closely followed by phase I outcomes, phase I costs, and phase II outcomes. The other inputs have less influence on the NPV, some hardly affecting it at all. This provides guidance about where to concentrate resources to have the greatest impact on the uncertainty around the NPV. As this model demonstrates, the use of logical statements are a simple, easy-to-comprehend, powerful, and often preferred way to incorporate dependencies between uncertain factors.

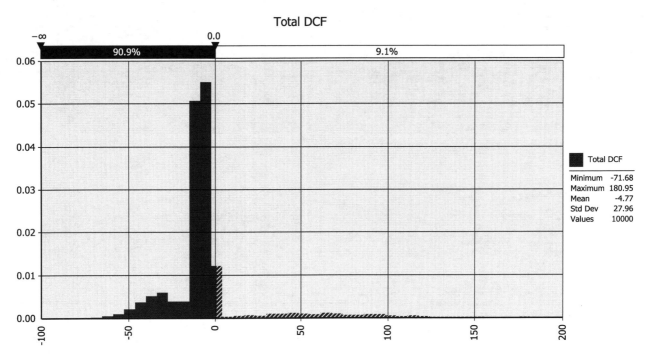

FIGURE 4.3
Drug Development Results

4.2 Second Example: Collateralized Debt Obligations

Collateralized Debt Obligations (CDOs) lie at the heart of the financial meltdown of 2007–2008. CDOs are asset-backed securities that are composed of a number of underlying assets[6] and are divided into different *tranches* having different risk characteristics.[7] Junior tranches default when one (or only a few) of the underlying assets default. Senior tranches only default when all of the underlying assets default. In this way, by combining a number of different risky assets, the senior tranches may be far less risky than the underlying assets. Individual tranches were then sold as single investments. Numerous CDOs were created, and senior tranches earned top ratings (for low risk) even when the underlying assets were each quite risky.[8] CDOs originated in 1987, but grew rapidly, especially after 2000. Unfortunately, many CDOs were backed by subprime mortgage bonds, and rising defaults in 2007 were a major contributing factor to the global financial crisis. Global CDO issuance reached over $520 billion in 2006, and, as a result of the financial crisis, fell to $6.4 billion in 2010.

A key factor determining the risk of a CDO is the degree of dependence of the underlying assets. Independence between the risks of all assets would permit many quite risky assets to be combined

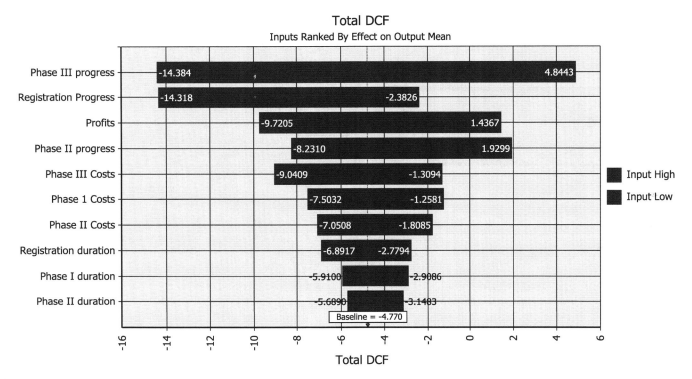

FIGURE 4.4
Tornado Chart for Drug Development Inputs

to produce senior tranches with a very low aggregate risk, since default would require simultaneous default of all of the underlying assets. This, of course, requires that the assets truly be independent. Conversely, if the underlying asset risks are highly correlated, then the senior tranche may not be much less risky than the individual assets. During the years before the crisis, it was thought that mortgage risks were geographically uncorrelated: default rates in one region were largely independent of default rates in other regions. This belief turned out to be incorrect. As one expert (Posner 2010) observes:

> However, if not used properly, Monte Carlo models can be hazardous to your health. Their results are unreliable if you do not accurately estimate the correlation between variables. These estimates require a sound grasp of underlying causative relationships. To those who miss this point, the models produce an illusion of precision, even though they are built upon the imprecision of random outcomes. Misuse of Monte Carlo models has contributed to some of the biggest Black Swan events of recent years, including the 1998 failure of Long-Term Capital Management (LTCM) and, just a few years later, the collapse of the multi-hundred billion dollar correlation trading market.[9]

To illustrate this further, we constructed a simplified model of two assets with risk characteristics of junk bond status (i.e., high-default risk). Two tranches are offered: the junior tranche defaults if *either* asset defaults, while the senior trance defaults only when *both* assets default. The spreadsheet CDO4.xlsx contains our model.[10] Annual default probabilities and a ratings schedule are shown in Columns G and H. The simulation of individual security default rates and their associated ratings, along with those for the CDOs, are shown in Figure 4.5.

Default for individual securities is modeled using Bernoulli distributions: we have chosen default rates of .0204 for each security, matching a BB− rating (just above junk bond status). After running the simulation, the mean default rate for each security appears in cells B4 and C4 and the associated ratings are provided using VLOOKUP functions. LOOKUP functions are a useful way of relating variables when the value a variable takes depends on where it lies within a range of possible values. In this case, the VLOOKUP function starts with the mean default rate, finds the lowest default rate in the ratings table that is larger,[11] and returns the value shown in the second column of the ratings table (the actual rating associated with that default rate).

Modeling the default rates for the CDOs is accomplished using logical IF statements. The junior tranche will default if either underlying asset defaults, while the senior tranche only defaults when both assets default.[12]

All that remains is to examine how different degrees of correlation between the two underlying assets can affect the resulting CDO default rates. We use a RiskSimTable to permit correlations ranging from 0 to 1 (modeled with a Gaussian copula, which is explained later in this chapter) between the two assets, as represented by a value shown in the @Risk Copula matrix, cell D28 (that ranges from 0 to 1).[13] We can think of the IF function as modeling a perfect correlation between variables: if A, then B. Correlations that are less than perfect are equivalent to saying something like, "if A, then sometimes B," or, "if A is higher, then B tends to be higher or lower." The frequency with which these events occur together reflects the degree to which they are correlated. For positive correlations, the linear correlation can range between 0 (independent events) and 1 (perfectly correlated events—these would be better modeled using IF statements).

For each potential correlation, we estimate the probability of default and associated rating for the junior and senior tranches in Columns C through F, Rows 12 through 22. We estimate this probability by

Modeling Relationships

	A	B	C	D	E	F	G	H
1	Individual Securities	1	2				annualized default rate	Fitch Ratings
2	Default Rate	=G15	=G15				0	AAA
3	Default?	=RiskBernoulli(B2,RiskCopula(Copula1,1))	=RiskBernoulli(C2,RiskCopula(Copula1,2))				0.0002	AAA
4	probability	=RiskMean(B3)	=RiskMean(C3)				0.0006	AA+
5	Rating	=VLOOKUP(B4,$G$2:$H$22,2,TRUE)	=VLOOKUP(C4,$G$2:$H$22,2,TRUE)				0.0009	AA
6							0.0012	AA-
7	Tranches	Default?	probability	Rating			0.0017	A+
8	Junior	=RiskOutput()+IF(B3+C3>=1,1,0)	=RiskMean(B8)	=VLOOKUP(C8,G3:H22,TRUE)			0.0019	A
9	Senior	=RiskOutput()+IF(B3+C3=2,1,0)	=RiskMean(B9)	=VLOOKUP(C9,G2:H22,TRUE)			0.0025	A-
10							0.0032	BBB+
11	simulation	Correlation	Junior tranche prob	Junior rating	Senior tranche prob	Senior rating	0.0038	BBB
12	1	0	=RiskMean($B$27,$A12)	=VLOOKUP(C12,$G$2:$H$22,2,TRUE)	=RiskMean(B$28,A12)	=VLOOKUP(E12,$G$2:$H$22,2,TRUE)	0.0075	BBB-
13	2	0.1	=RiskMean($B$27,$A13)	=VLOOKUP(C13,$G$2:$H$22,2,TRUE)	=RiskMean(B$28,A13)	=VLOOKUP(E13,$G$2:$H$22,2,TRUE)	0.0107	BB+
14	3	0.2	=RiskMean($B$27,$A14)	=VLOOKUP(C14,$G$2:$H$22,2,TRUE)	=RiskMean(B$28,A14)	=VLOOKUP(E14,$G$2:$H$22,2,TRUE)	0.0145	BB
15	4	0.3	=RiskMean($B$27,$A15)	=VLOOKUP(C15,$G$2:$H$22,2,TRUE)	=RiskMean(B$28,A15)	=VLOOKUP(E15,$G$2:$H$22,2,TRUE)	0.0204	BB-
16	5	0.4	=RiskMean($B$27,$A16)	=VLOOKUP(C16,$G$2:$H$22,2,TRUE)	=RiskMean(B$28,A16)	=VLOOKUP(E16,$G$2:$H$22,2,TRUE)	0.0259	B+
17	6	0.5	=RiskMean($B$27,$A17)	=VLOOKUP(C17,$G$2:$H$22,2,TRUE)	=RiskMean(B$28,A17)	=VLOOKUP(E17,$G$2:$H$22,2,TRUE)	0.0324	B
18	7	0.6	=RiskMean($B$27,$A18)	=VLOOKUP(C18,$G$2:$H$22,2,TRUE)	=RiskMean(B$28,A18)	=VLOOKUP(E18,$G$2:$H$22,2,TRUE)	0.043	B-
19	8	0.7	=RiskMean($B$27,$A19)	=VLOOKUP(C19,$G$2:$H$22,2,TRUE)	=RiskMean(B$28,A19)	=VLOOKUP(E19,$G$2:$H$22,2,TRUE)	0.0568	CCC+
20	9	0.8	=RiskMean($B$27,$A20)	=VLOOKUP(C20,$G$2:$H$22,2,TRUE)	=RiskMean(B$28,A20)	=VLOOKUP(E20,$G$2:$H$22,2,TRUE)	0.0664	CCC
21	10	0.9	=RiskMean($B$27,$A21)	=VLOOKUP(C21,$G$2:$H$22,2,TRUE)	=RiskMean(B$28,A21)	=VLOOKUP(E21,$G$2:$H$22,2,TRUE)	0.147	CC
22	11	1	=RiskMean($B$27,$A22)	=VLOOKUP(C22,$G$2:$H$22,2,TRUE)	=RiskMean(B$28,A22)	=VLOOKUP(E22,$G$2:$H$22,2,TRUE)	0.2996	C
23								
24	This Simulation	=RiskSimtable(B12:B22)						
25	Simulating Copula		@RISK Copula: Copula1					
26			Type		Gaussian			
27	Junior Tranche	=IF(B3+C3>=1,1,0)	Matrix	1				
28	Senior Tranche	=IF(B3+C3=2,1,0)		=B24	1			

FIGURE 4.5
CDO Model

looking at how often it defaults and using the RiskMean to estimate the fraction of time that this occurs. The correlation is captured in the model in cells B3 and C3 by adding the RiskCopula to the RiskBernoulli distributions. The addition of this argument to the function is what tells @Risk to use the correlation when simulating the default distributions, and the use of the RiskSimTable in cell B24 instructs @Risk to vary the correlations from 0 to 1.

The final step is to model the default probabilities of the tranches using the listed correlations between the assets. In @Risk, this is accomplished by adding the second argument, indicating the simulation number, in the probabilities of junior and senior tranche defaults (C12:C22 and E12:E22).

Running 11 simulations of 10,000 iterations each provides the results shown in Figure 4.6.

The table in Figure 4.6 shows that the junior tranche ratings (in Column D) are not much affected by correlation between the assets, and are at, or just above, junk bond status. The probability of default of the junior tranche does double, though, from about 2% to 4%

	A	B	C	D	E	F
11	simulation	Correlation	Junior tranche prob	Junior rating	Senior tranche	Senior rating
12	1	0	0.04040	B	0.00040	AAA
13	2	0.1	0.04050	B	0.00030	AAA
14	3	0.2	0.04010	B	0.00070	AA+
15	4	0.3	0.03950	B	0.00130	AA-
16	5	0.4	0.03900	B	0.00180	A+
17	6	0.5	0.03740	B	0.00340	BBB+
18	7	0.6	0.03680	B	0.00400	BBB
19	8	0.7	0.03380	B	0.00700	BBB
20	9	0.8	0.03230	B+	0.00850	BBB-
21	10	0.9	0.02960	B+	0.01120	BB+
22	11	1	0.02040	BB-	0.02040	BB-

FIGURE 4.6
CDO Model Results

(Column C). However, the senior tranche ratings (Column F) are quite affected by correlation: when there is little correlation between the two underlying assets (shown in Rows 12 and 13), the CDO has a AAA rating, but when the two assets are perfectly correlated (shown in Row 22), then the CDO senior tranche is rated BB–. The protection of the senior tranche comes from the requirement that both assets must default before the senior tranche holders are liable—therefore, when the assets are perfectly correlated, there is no extra protection compared to the junior tranche. But when the assets are uncorrelated, then bundling high-risk assets can produce a senior tranche CDO that is very high quality (low risk). A contributing factor to the financial meltdown of the late 2000s was the belief that the underlying assets in these CDOs were indeed uncorrelated, when in fact, they were not. That belief was based, in turn, on the belief that real estate markets were essentially uncorrelated so that pooling mortgages from different parts of the country effectively reduced the risk of simultaneous default.

4.3 Multiple Correlations Example: Cockpit Failures

In order to include relationships between (many) more than two variables in a simulation model, using an approach where we have to manually correlate each variable with every other (for example, as with IF statements) could take a lot of time. For example, with six variables, there are $5+4+3+2+1=15$ relationships. In those cases of many-to-many correlations, it can often be useful to build and use a *correlation matrix* that reflects the correlations among the various

	J	K	L	M	N	O	P	Q	R	S	T	U	V	W	X	Y	Z	AA	AB	AC	AD	AE	AF	AG	AH	AI	AJ	AK	AL	AM	AN	AO	AP	AQ	AR	AS
1	@RISK Corre	ADF Antenna	ADF Display	ADF Receiver	Airspeed Ind	Airspeed Ind	Alternator / T	Altimeter / T	Ammeter Vac	Attitude Indi	Clock / Time	Comm Anten	Comm Radio	Directional G	Fuel Quantity	GS Antenna /	ILS Display /	Left Fuel Qua	LOC Antenna	Magnetic Cor	Marker Beac	Marker Beac	Oil Temp Pre	Pilot System	Receiver / Ti	Right Fuel Q	Suction Gaug	Tachometer	Transponder	Transponder	Turn Coordin	Vacuum Pum	Vertical Spee	VOR Antenna	VOR Display	VOR Receive
2	ADF Antenna	1
3	ADF Display	0.8	1
4	ADF Receiver	0.8	0.8	1
5	Airspeed Indi	0.8	0.8	0.8	1
6	Airspeed Indi	0.8	0.8	0.8	0.8	1
7	Alternator / V	0.8	0.8	0.8	0.8	0.8	1
8	Altimeter / T	0.8	0.8	0.8	0.8	0.8	0.8	1
9	Ammeter Vac	0.8	0.8	0.8	0.8	0.8	0.8	0.8	1
10	Attitude Indi	0.8	0.8	0.8	0.8	0.8	0.8	0.8	0.8	1
11	Clock / Time	0.8	0.8	0.8	0.8	0.8	0.8	0.8	0.8	0.8	1
12	Comm Anten	0.8	0.8	0.8	0.8	0.8	0.8	0.8	0.8	0.8	0.8	1
13	Comm Radio	0.8	0.8	0.8	0.8	0.8	0.8	0.8	0.8	0.8	0.8	0.8	1
14	Directional G	0.8	0.8	0.8	0.8	0.8	0.8	0.8	0.8	0.8	0.8	0.8	0.8	1
15	Fuel Quantity	0.8	0.8	0.8	0.8	0.8	0.8	0.8	0.8	0.8	0.8	0.8	0.8	0.8	1
16	GS Antenna /	0.8	0.8	0.8	0.8	0.8	0.8	0.8	0.8	0.8	0.8	0.8	0.8	0.8	0.8	1
17	ILS Display /	0.8	0.8	0.8	0.8	0.8	0.8	0.8	0.8	0.8	0.8	0.8	0.8	0.8	0.8	0.8	1
18	Left Fuel Qua	0.8	0.8	0.8	0.8	0.8	0.8	0.8	0.8	0.8	0.8	0.8	0.8	0.8	0.8	0.8	0.8	1
19	LOC Antenna	0.8	0.8	0.8	0.8	0.8	0.8	0.8	0.8	0.8	0.8	0.8	0.8	0.8	0.8	0.8	0.8	0.8	1
20	Magnetic Cor	0.8	0.8	0.8	0.8	0.8	0.8	0.8	0.8	0.8	0.8	0.8	0.8	0.8	0.8	0.8	0.8	0.8	0.8	1
21	Marker Beac	0.8	0.8	0.8	0.8	0.8	0.8	0.8	0.8	0.8	0.8	0.8	0.8	0.8	0.8	0.8	0.8	0.8	0.8	0.8	1
22	Marker Beac	0.8	0.8	0.8	0.8	0.8	0.8	0.8	0.8	0.8	0.8	0.8	0.8	0.8	0.8	0.8	0.8	0.8	0.8	0.8	0.8	1
23	Oil Temp Pre	0.8	0.8	0.8	0.8	0.8	0.8	0.8	0.8	0.8	0.8	0.8	0.8	0.8	0.8	0.8	0.8	0.8	0.8	0.8	0.8	0.8	1
24	Pilot System	0.8	0.8	0.8	0.8	0.8	0.8	0.8	0.8	0.8	0.8	0.8	0.8	0.8	0.8	0.8	0.8	0.8	0.8	0.8	0.8	0.8	0.8	1
25	Receiver / Ti	0.8	0.8	0.8	0.8	0.8	0.8	0.8	0.8	0.8	0.8	0.8	0.8	0.8	0.8	0.8	0.8	0.8	0.8	0.8	0.8	0.8	0.8	0.8	1
26	Right Fuel Q	0.8	0.8	0.8	0.8	0.8	0.8	0.8	0.8	0.8	0.8	0.8	0.8	0.8	0.8	0.8	0.8	0.8	0.8	0.8	0.8	0.8	0.8	0.8	0.8	1
27	Suction Gaug	0.8	0.8	0.8	0.8	0.8	0.8	0.8	0.8	0.8	0.8	0.8	0.8	0.8	0.8	0.8	0.8	0.8	0.8	0.8	0.8	0.8	0.8	0.8	0.8	0.8	1
28	Tachometer	0.8	0.8	0.8	0.8	0.8	0.8	0.8	0.8	0.8	0.8	0.8	0.8	0.8	0.8	0.8	0.8	0.8	0.8	0.8	0.8	0.8	0.8	0.8	0.8	0.8	0.8	1
29	Transponder	0.8	0.8	0.8	0.8	0.8	0.8	0.8	0.8	0.8	0.8	0.8	0.8	0.8	0.8	0.8	0.8	0.8	0.8	0.8	0.8	0.8	0.8	0.8	0.8	0.8	0.8	0.8	1
30	Transponder	0.8	0.8	0.8	0.8	0.8	0.8	0.8	0.8	0.8	0.8	0.8	0.8	0.8	0.8	0.8	0.8	0.8	0.8	0.8	0.8	0.8	0.8	0.8	0.8	0.8	0.8	0.8	0.8	1
31	Turn Coordin	0.8	0.8	0.8	0.8	0.8	0.8	0.8	0.8	0.8	0.8	0.8	0.8	0.8	0.8	0.8	0.8	0.8	0.8	0.8	0.8	0.8	0.8	0.8	0.8	0.8	0.8	0.8	0.8	0.8	1
32	Vacuum Pum	0.8	0.8	0.8	0.8	0.8	0.8	0.8	0.8	0.8	0.8	0.8	0.8	0.8	0.8	0.8	0.8	0.8	0.8	0.8	0.8	0.8	0.8	0.8	0.8	0.8	0.8	0.8	0.8	0.8	0.8	1
33	Vertical Spee	0.8	0.8	0.8	0.8	0.8	0.8	0.8	0.8	0.8	0.8	0.8	0.8	0.8	0.8	0.8	0.8	0.8	0.8	0.8	0.8	0.8	0.8	0.8	0.8	0.8	0.8	0.8	0.8	0.8	0.8	0.8	1
34	VOR Antenna	0.8	0.8	0.8	0.8	0.8	0.8	0.8	0.8	0.8	0.8	0.8	0.8	0.8	0.8	0.8	0.8	0.8	0.8	0.8	0.8	0.8	0.8	0.8	0.8	0.8	0.8	0.8	0.8	0.8	0.8	0.8	0.8	1
35	VOR Display	0.8	0.8	0.8	0.8	0.8	0.8	0.8	0.8	0.8	0.8	0.8	0.8	0.8	0.8	0.8	0.8	0.8	0.8	0.8	0.8	0.8	0.8	0.8	0.8	0.8	0.8	0.8	0.8	0.8	0.8	0.8	0.8	0.8	1
36	VOR Receive	0.8	0.8	0.8	0.8	0.8	0.8	0.8	0.8	0.8	0.8	0.8	0.8	0.8	0.8	0.8	0.8	0.8	0.8	0.8	0.8	0.8	0.8	0.8	0.8	0.8	0.8	0.8	0.8	0.8	0.8	0.8	0.8	0.8	0.8	1

FIGURE 4.7
Correlation Matrix

inputs. We return to our example of cockpit system failures from Chapter 3 to illustrate this. Figure 4.7 shows a possible correlation matrix for the 35 cockpit component failures, with assumed correlation of 0.8 for each pair of components.

Each cell (all 595 of them; $34 + 33 + 32 + \cdots + 1$) of the matrix shows an assumed correlation between a pair of components. The values of 1 on the diagonal reflect the fact that each component is perfectly correlated with itself. Only one side of the diagonal needs to be filled out, since the matrix is symmetric. This matrix was created using "Define Correlation Matrix" functionality, found under Define Correlations on the @Risk toolbar. In the dialog, all of the individual components were added as inputs to form the matrix, as shown in Figure 4.8.

With this many correlations, it is entirely possible that, when we enter the degree of correlation (ranging from –1 to +1)[14] for all (595) of them, there may be inconsistencies. For example, we might assume that A and C are not correlated with each other but have positive correlations of *each* with B: this would therefore require that A and C also be correlated. To check that the correlations are consistent, and to correct them if they are not, we can use the RiskCorrectCormat function (found under Insert Function—Miscellaneous on the @Risk toolbar), and selecting the entire correlation matrix. In this case, the correlation matrix does not need correcting, since we have assumed that the correlations between all 35 variables are equal to 0.8. In reality, we would probably have various subsystems with differing

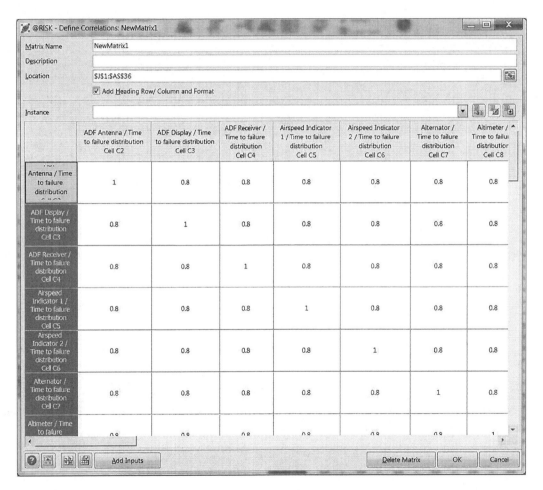

FIGURE 4.8
Creation of the Correlation Matrix

Modeling Relationships

	A	B	C	D	E	F
1	Event (system)	MTBF (hours)	Time to failure distribution		first failure	=RiskOutput()+MIN(C2:C36)
2	ADF Antenna	4000	=RiskExpon(B2,RiskCorrmat(NewMatrix1,1))		maintenance interval	failure probability
3	ADF Display	19900	=RiskExpon(B3,RiskCorrmat(NewMatrix1,2))	1		=RiskTarget($F$1,E3)
4	ADF Receiver	4200	=RiskExpon(B4,RiskCorrmat(NewMatrix1,3))	2		=RiskTarget($F$1,E4)
5	Airspeed Indicator 1	18100	=RiskExpon(B5,RiskCorrmat(NewMatrix1,4))	3		=RiskTarget($F$1,E5)
6	Airspeed Indicator 2	2500	=RiskExpon(B6,RiskCorrmat(NewMatrix1,5))	4		=RiskTarget($F$1,E6)
7	Alternator	7600	=RiskExpon(B7,RiskCorrmat(NewMatrix1,6))	5		=RiskTarget($F$1,E7)
8	Altimeter	5500	=RiskExpon(B8,RiskCorrmat(NewMatrix1,7))	6		=RiskTarget($F$1,E8)
9	Ammeter Vacuum Pressure	21500	=RiskExpon(B9,RiskCorrmat(NewMatrix1,8))	7		=RiskTarget($F$1,E9)
10	Attitude Indicator	2500	=RiskExpon(B10,RiskCorrmat(NewMatrix1,9))	8		=RiskTarget($F$1,E10)
11	Clock	17600	=RiskExpon(B11,RiskCorrmat(NewMatrix1,10))	9		=RiskTarget($F$1,E11)
12	Comm Antenna	1200	=RiskExpon(B12,RiskCorrmat(NewMatrix1,11))	10		=RiskTarget($F$1,E12)
13	Comm Radio	900	=RiskExpon(B13,RiskCorrmat(NewMatrix1,12))	11		=RiskTarget($F$1,E13)
14	Directional Gyro	3400	=RiskExpon(B14,RiskCorrmat(NewMatrix1,13))	12		=RiskTarget($F$1,E14)
15	Fuel Quantity Indicator	16500	=RiskExpon(B15,RiskCorrmat(NewMatrix1,14))	13		=RiskTarget($F$1,E15)
16	GS Antenna	900	=RiskExpon(B16,RiskCorrmat(NewMatrix1,15))	14		=RiskTarget($F$1,E16)
17	ILS Display	10000	=RiskExpon(B17,RiskCorrmat(NewMatrix1,16))	15		=RiskTarget($F$1,E17)
18	Left Fuel Quantity Transduce	51400	=RiskExpon(B18,RiskCorrmat(NewMatrix1,17))	16		=RiskTarget($F$1,E18)
19	LOC Antenna	900	=RiskExpon(B19,RiskCorrmat(NewMatrix1,18))	17		=RiskTarget($F$1,E19)
20	Magnetic Comp	19900	=RiskExpon(B20,RiskCorrmat(NewMatrix1,19))	18		=RiskTarget($F$1,E20)
21	Marker Beacon Antenna	14800	=RiskExpon(B21,RiskCorrmat(NewMatrix1,20))	19		=RiskTarget($F$1,E21)
22	Marker Beacon Receiver	5300	=RiskExpon(B22,RiskCorrmat(NewMatrix1,21))	20		=RiskTarget($F$1,E22)
23	Oil Temp Pressure Gauge	6200	=RiskExpon(B23,RiskCorrmat(NewMatrix1,22))	21		=RiskTarget($F$1,E23)
24	Pilot System	73600	=RiskExpon(B24,RiskCorrmat(NewMatrix1,23))	22		=RiskTarget($F$1,E24)
25	Receiver	10000	=RiskExpon(B25,RiskCorrmat(NewMatrix1,24))	23		=RiskTarget($F$1,E25)
26	Right Fuel Quantity Transduce	51400	=RiskExpon(B26,RiskCorrmat(NewMatrix1,25))	24		=RiskTarget($F$1,E26)
27	Suction Gauge	21500	=RiskExpon(B27,RiskCorrmat(NewMatrix1,26))			
28	Tachometer	8400	=RiskExpon(B28,RiskCorrmat(NewMatrix1,27))			

FIGURE 4.9
Completed Cockpit Model with Correlations

correlations between their components, and the correlation matrix would be much more complex (and may become inconsistent with this many components). Alternatively, if we had access to relevant data, we could possibly use historical failure data to estimate the relationships between different failure modes.

When the (valid) correlation matrix is entered into the spreadsheet, the correlations are captured in each of the input distributions, as shown in Figure 4.9 (see Column C, with the RiskCormat function included in each of the 35 Exponential probability distributions).

We have retained the table (cells E1:F26) showing the failure probabilities within the first 1 through 24 hours. The results of the simulation are shown in Figure 4.10, where we have pasted the results from the zero correlation case (from Chapter 3) into Column G.

	E	F	G
2	maintenance interval	failure probability	w/o correlation
3	1	0.49%	0.97%
4	2	0.89%	1.93%
5	3	1.20%	2.91%
6	4	1.54%	3.93%
7	5	1.81%	4.89%
8	6	2.11%	5.82%
9	7	2.34%	6.73%
10	8	2.64%	7.69%
11	9	2.93%	8.66%
12	10	3.15%	9.52%
13	11	3.41%	10.36%
14	12	3.80%	11.31%
15	13	4.03%	12.17%
16	14	4.31%	13.03%
17	15	4.62%	13.88%
18	16	4.95%	14.78%
19	17	5.22%	15.62%
20	18	5.49%	16.36%
21	19	5.75%	17.18%
22	20	5.99%	18.01%
23	21	6.24%	18.83%
24	22	6.46%	19.54%
25	23	6.71%	20.37%
26	24	7.07%	21.12%

FIGURE 4.10
Cockpit Failure Simulation Results: With and Without Correlation

We have labeled the hourly failure probabilities as "maintenance intervals" since maintenance decisions will largely be based on the likelihood of failures within those time frames. As the results show, a positive correlation between failures of components clearly reduces the failure probabilities within the 24-hour interval. The effect of positive correlations in a simulation model can therefore sometimes increase the risk, and sometimes decrease it. Ignoring correlations can therefore result in substantial errors in our estimates of relevant risks.

4.4 Copulas Example: How Correlated Are Home Prices?

So, just how correlated are home prices (recall that presumed lack of correlation is what made CDOs less risky)? CityHomePrices4.xlsx provides monthly data on home prices, and changes in home prices, from 1990 through 2018 in two American cities, Seattle (in Washington state) and Las Vegas (in Nevada)—Las Vegas was at the heart of the subprime mortgage crisis.[15] We wish to investigate the correlation patterns. Figure 4.11 is an index of home prices in the two cities (January 1, 2000, is the base, set equal to 100), and shows how home prices have varied over the time period.

It is clear from Figure 4.11 that there is a relationship (a correlation) between the two variables. It is also evident that there is a correlation *within* each series—this is called *autocorrelation*, but we will delay examining this type of correlation until the next chapter. Figure 4.12 shows the month-to-month price changes.

Again, the correlation is clear, as is the autocorrelation (perhaps less evident visually, but we will again postpone consideration of

Modeling Relationships

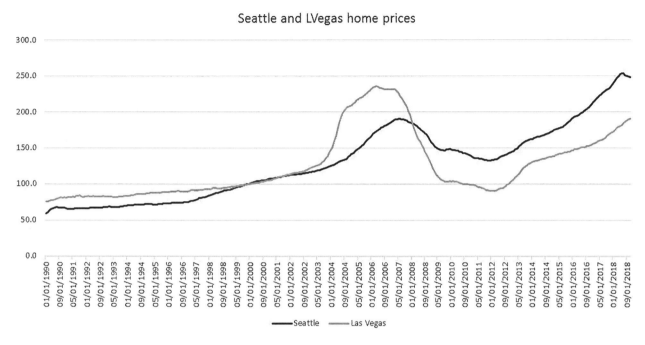

FIGURE 4.11
Seattle and Las Vegas Home Prices

this until Chapter 5). Ignoring the time series nature of this data, but focusing on the correlation between the monthly changes of price in the two cities, we want to examine the relationship between the two monthly price changes, shown in Figure 4.13.

There appears to be a clear pattern of positive correlation, but the correlation is tighter at the low ends of the distributions and weaker in the middle and upper ends.[16] A Gaussian copula (which produces a symmetric and "linear" correlation) will not capture this pattern well. It is widely believed (at least now) that housing markets are more tightly coupled when prices are falling significantly than when the changes are more moderate.[17]

There are several different copula structures available in the finance literature and in @Risk. Among these are correlation patterns (copulas) that can better represent this pattern than does the Gaussian copula.

FIGURE 4.12
Monthly Home Price Changes in Seattle and Las Vegas

These patterns can be estimated from data or chosen by the analyst, just as with probability distributions (see Chapter 3). @Risk permits a number of copulas to be fit to data—these are found under the Define Correlations button on the @Risk toolbar. Selecting Fit Copula and using all eligible ("candidate") copula functions gives us Figure 4.14.

Based on the ranking by the information criteria (in this case, we used the Akaike information criterion, i.e., the AIC[18]), the Clayton copula fits this data best, followed by the Gumbel copula. The pattern of the data (and the Clayton copula fit) shows a fairly strong ("tight") correlation at the largest (in absolute value) home price decreases, with a more diffuse pattern when prices are rising.

@Risk then gives the choice of placing the copula into the spreadsheet directly or attaching it to Distribution functions. The latter would be appropriate if we fit distribution functions to each

Modeling Relationships

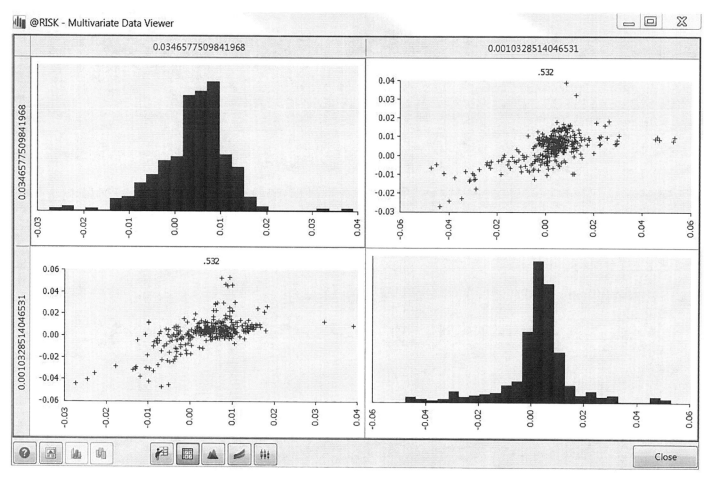

FIGURE 4.13
Scatterplot of Monthly Home Price Changes

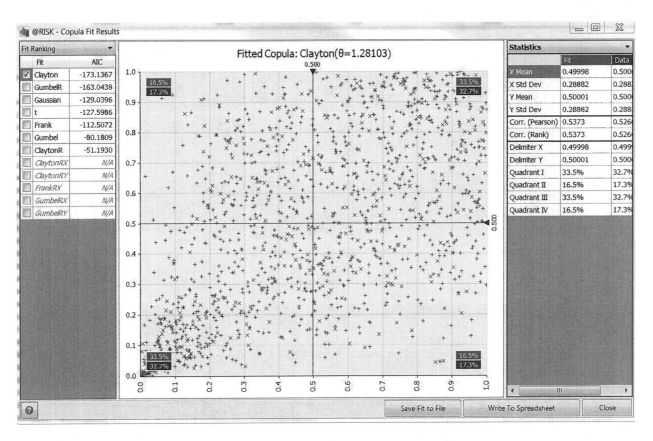

FIGURE 4.14
Copula Fit for Home Price Change Data

of the individual city home price changes and then wish to use the copula to capture the correlation between them. If we want to model the city home prices later, we can just save the copula and then attach it to the simulated distributions by adding the RiskCopula as an additional entry in the distribution function, as shown in Figure 4.15.

If we had attached the copula directly to the distribution inputs, then the last arguments in the functions, RiskCopula(Copula1,#), where # indicates each of the variables to be correlated, would have been added for us, rather than adding them manually. The Laplace distributions fitted each of the monthly price-change data best, and this model can now be used to examine and simulate house price changes (and their relationships) in the two cities.

It is also possible to model the correlation patterns among more than two variables. We have added the monthly home price changes for Boston into CityHomePrices4.xlsx and fitted various copulas to the three cities simultaneously. Again, the best fitting copula is the Clayton, as shown in Figure 4.16.

Given the role played by correlation in home price changes in the Great Recession, it is worth asking whether that correlation could have been foreseen. The 1990s and early 2000s were an era of generally rising home prices, so perhaps the asymmetric correlation pattern was not evident at the time. The copula fit for the 1990–2006 time frame (the recession began in December 2007) also shows the Clayton copula as the best fit. However, it is a weaker correlation pattern than that produced using the entire time period data.

4.5 Empirical Copulas

There is an additional type of copula that is available in @Risk: an empirical copula. This copula will mimic whatever pattern the data reflects—even highly complex nonlinear relationships that could not otherwise be captured by the other "parametric" copula patterns. Empirical4.xlsx contains hypothetical data illustrating an unusual correlation pattern. Figure 4.17 shows a scatterplot of the data.

The correlation pattern is clearly not typically encountered. We can fit distributions to the two variables (the Laplace for Variable

	J	K
1	@RISK Copula: Copula1	
2	Type	Clayton
3	Dimension	2
4	Parameter	1.28102723530577
5		
6	Seattle	=RiskLaplace(0.004988,0.0077634,RiskName("Dataset 1"),RiskCopula(Copula1,1))
7	Las Vegas	=RiskLaplace(0.0037985,0.011294,RiskName("Dataset 2"),RiskCopula(Copula1,2))

FIGURE 4.15
Adding a Copula to Distribution Inputs

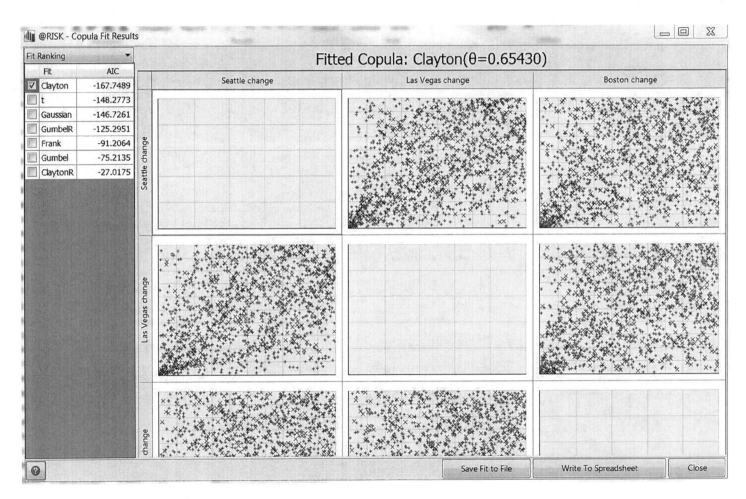

FIGURE 4.16
Copula Fit for Three Cities' Home Price Changes

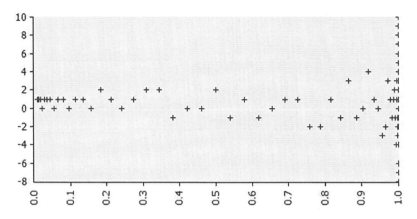

FIGURE 4.17
Hypothetical Correlation Pattern

1 and the Triangular for Variable 2), and we also fit a copula to the two series (the ClaytonR fit best). To fit an Empirical copula—a non-parametric function to approximate the correlation pattern found in the data—we use Define Correlations—Define Copula and select the Empirical copula and deselect all the others. The spreadsheet section for the simulation of the two variables is shown in Figure 4.18.

Running the simulation, and selecting the scatterplot output for the two variables (the scatterplot is found at the bottom of the @Risk Results window), we get the two scatterplots shown in Figure 4.19, one based on the best-fitting "parametric" copula and the other using the empirical "non-parametric" copula.

Both scatterplots—using the ClaytonR and Empirical copulas—are similar and seem to reasonably approximate the original data. The Empirical copula seems to visually match the pattern in Figure 4.17 a bit more closely.[19] So, empirical copulas provide an alternative way to model correlation patterns that may not be amenable to parametric copula patterns.

The Empirical copula is always an option for modeling correlation patterns (either in bivariate or multivariate data). The advantage is that it will more closely match the existing empirical data than any other "parametric" copula. When using empirical copulas, there are, however, two shortcomings to recognize. First, empirical copulas are computationally intensive, so extensive use of them can considerably slow down simulation models. Second, and more importantly, it may not always be desirable to only match the pattern in the data. After all, closely matching past data may be a poor way to forecast the future. It may be more accurate to represent the overall pattern in the data (and include some extrapolation and interpolation) than to

	D	E	F	G	H
1	Var1	=RiskLaplace(1,3.5214,RiskName("Variable 1"),RiskCopula(Copula1,1))		V1-empirical	=RiskLaplace(1,3.5214,RiskName("V1-empirical"),RiskCopula(Copula2,1))
2	Var2	=RiskTriang(-0.12932,1,1,RiskName("Variable 2"),RiskCopula(Copula1,2))		V2-empirical	=RiskTriang(-0.12932,1,1,RiskName("V2-empirical"),RiskCopula(Copula2,2))
3					
4	@RISK Copula: Copula1			@RISK Copula: Copula2	
5	Type	ClaytonR		Type	Empirical
6	Dimension	2		Interpolate	TRUE
7	Parameter	0.391607715514558		Source Data	=RiskEmpiricalCopulaData($A$2:$B$101)

FIGURE 4.18
Simulation Model for the Hypothetical Data

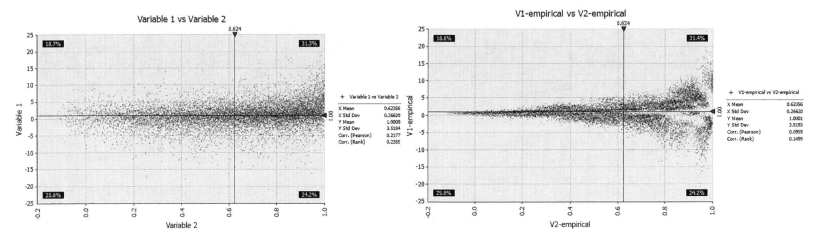

FIGURE 4.19
Simulation Results Using Parametric and Empirical Copulas

match one particular realization of the unobservable process that generated the data.

4.6 Fifth Example: Advertising Effectiveness

Suppose a company has data on levels of advertising ($ thousands/month) and sales ($ thousands/month) from 10 different local markets. The data is given in Table 4.1 (and Regression Model4.xlsx).

The company wishes to estimate the effectiveness of their advertising and to predict the level of sales for a certain advertising budget. Taking a very simplistic (and, as we will see, incorrect) approach,

TABLE 4.1

Advertising Effectiveness

Advertising	Sales
1	2
2	7
3	3
4	7
5	10
6	9
7	13
8	12
9	19
10	18

we could attempt to model the relationship between advertising and sales using a LOOKUP function. For an advertising budget of $7500, this will produce estimated sales of $13,000 (if we don't look for an exact match—an exact match will produce an Excel error since there is no exact match in the table). This LOOKUP approach, however, ignores any uncertainties and does not interpolate between different advertising budgets.

An alternative approach might be to estimate a copula to capture the relationship between the two variables. But this approach treats the two variables as randomly fluctuating, albeit with some dependence between them. While modeling a relationship using a copula approach does capture the uncertainty around the relationship, it assumes that both variables (i.e., advertising and sales) are uncertain. When one of the variables is not uncertain (in this case advertising) or is based on another number of factors (e.g., another part of the model), a copula approach is typically not appropriate. In this case, we can estimate the "dependency" relationship using a regression (in this case, a *simple linear regression analysis*).

There are many forms of regression analysis, and Excel is capable of producing a variety of these—we will only consider linear regression here. Linear regression assumes that there is a linear relationship between a response variable, Y, and a number of explanatory variables, $X_1,...,X_n$. In the present case, Y is sales, and we only have one X variable, advertising (i.e., a simple linear regression).[20]

4.7 Regression Modeling

In this section and the next, we will discuss how (linear) regressions can also be used to include relationships in simulation models. While we provide some illustrative examples, we want to emphasize to the reader that considerations that go beyond the scope of our text could go into performing valid regression (e.g., collinearity, confounding). The purpose of this section and the next is therefore *not* to give a thorough thesis of how to perform regressions, but instead to illustrate how such analyses can be used in and augmented with using simulation models.

Linear regression does impose a few restrictions on the relationship between Y and X. The assumptions that underlie linear regression models are[21]:

- The data comes from a sufficiently large *random* sample.
- A linear relationship exists between the Y variable and the X variable (or multiple X variables when multiple linear regression is used).
- The residuals (the difference between what the model predicts and the actual Y values) are normally distributed.
- The distributions of these residuals are identical and independent for different values of the X variables.

Regression models can sometimes be quite robust to deviations from these assumptions, however, and are commonly used to capture relationships from real data that does not strictly satisfy these conditions. Also, refinements of the standard regression model are available to deal with many common departures from these assumptions.[22]

Simple linear regression models are based on finding a well (or "best")-fitting linear model between the Y and X variables. "Best fitting" is often defined as the linear model that minimizes the sum of the squared deviations of the actual data points from the estimated linear relationship. Least-squares regression can therefore be thought of by considering many potential straight lines that can be drawn through the data points and finding the one that minimizes the sum of the squared deviations ("errors") of the points from each line. Figure 4.20 illustrates the procedure for our hypothetical advertising–sales data[23].

The arrow labeled "unexplained" in Figure 4.20 represents the deviation of that particular data point (advertising = 9, sales = 19) from the straight line $Y = 1.76X + 0.33$ which is the least-squares regression line (i.e., the "best-fitting line") for this data. The arrow labeled "explained" represents the deviation of this data point from the average sales value in the data (average sales = 10 in this dataset)—it is explained because our model assumes that sales are *dependent* on the advertising level, and the advertising level at that point is greater than the average value in the data (9 > 5.4). Since the advertising level at this point is relatively high, our least-squares regression model predicts (or explains) that the resulting sales will be higher than average (predicting a sales level of 16). It does not predict, however, why the sales are as high as 19. The additional $3000 of monthly sales ($19,000 minus $16,000) is "unexplained" by the linear regression model. It is the sum of all these unexplained squared deviations ("errors") that the least-squares regression line minimizes.

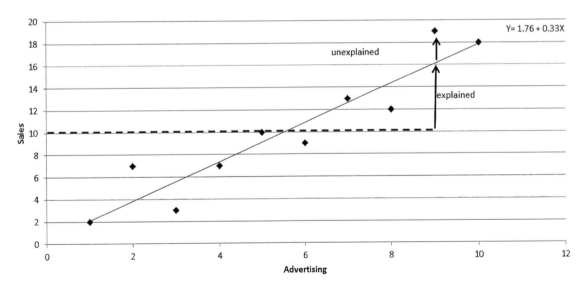

FIGURE 4.20
Advertising–Sales Regression Line

Excel provides a number of ways to obtain the least-squares regression line. A comprehensive overview of the output and results from a regression analysis in Excel can be obtained by choosing Data Analysis on the Data menu.[24] A variety of statistical methods are available within Data Analysis—choose Regression from the drop-down list. Insert the Sales data as the Y range and the Advertising data as the X range (also, click Labels and include the labels in the ranges), and you will get the output shown in Figure 4.21.

The Regression Statistics section provides some overall information and statistics about the relationship between X and Y.

Multiple R is a number between 0 and +1 that measures the quality of the linear regression (for a simple linear regression, this is the same as [the absolute value of] the Pearson correlation coefficient, which you can calculate using Excel's CORREL function). *R squared*, its squared value, measures the extent to which the independent variables account for the variation in the response variable. R squared ranges from 0 to 1: a value of 1 means that every actual value of Y is exactly equal to its predicted value. Thus, an R squared of 1 is equivalent to the least-squares regression line passing exactly through every actual data point. The *Adjusted R squared* makes an adjustment for the number of independent ("explanatory") variables (in this case just one, the advertising level): R squared will always increase if more independent variables are added to the model, but the Adjusted R squared will only increase when statistically "meaningful" independent variables are included. The *Standard Error* is the standard deviation of the residuals from the model (we'll discuss and use this metric more in the next section). *Observations* are the number of data points.

The ANOVA (analysis of variance) table also provides measures of the overall fit of the model to the data. It does not, however, add much to the interpretation of the model beyond what is in the other sections of the regression output. Briefly, it decomposes the *Total* sum of squared residuals in the regression into the fraction that is accounted for by the regression model (*Regression*) and that which is not (*Residual*). It divides these sums of squared residuals (*SS*) by the appropriate degrees of freedom (*df*—a function of the number of data points and the number of independent variables), to get the mean square (*MS*), and then gives the probability of getting a mean square ratio larger than this value if there was no relationship between the two variables, using an *F* distribution. The resulting *Significance F* is often interpreted as the probability of obtaining an F statistic at least as large as the one observed, given the null hypothesis of no dependency between advertising and sales is true (this interpretation is, strictly speaking, not correct, though). If

SUMMARY OUTPUT

Regression Statistics

Multiple R	0.93743687
R Square	0.87878788
Adjusted R Square	0.86363636
Standard Error	2.09617256
Observations	10

ANOVA

	df	SS	MS	F	Significance F
Regression	1	254.8484848	254.8485	58	6.21252E-05
Residual	8	35.15151515	4.393939		
Total	9	290			

	Coefficients	Standard Error	t Stat	P-value	Lower 95%	Upper 95%
Intercept	0.33333333	1.431958467	0.232781	0.8217762	-2.96876881	3.63543548
advertising	1.75757576	0.230781003	7.615773	6.213E-05	1.22539381	2.2897577

FIGURE 4.21
Excel Regression Output

there were no relationship between the independent and dependent variables, then the regression sum of squares would be likely be close to zero, so the MS and F values would be low, and the idea is that the Significance F would in that case only rarely (1 in 20 times) be <5%. Conversely, the greater the dependency, the larger the regression SS relative to the Residual SS, the greater the F value, and the smaller the Significance F. If this value is sufficiently low, then the probability that the independent variables could account for this much of the variation in the dependent variable is low enough to conclude that the overall relationship is significant and not just the result of chance.

The last table in Figure 4.21 provides information about the variables used in the model. The *Intercept* is part of the linear relationship, but it represents all of the variables that were not included in the model, so it is not generally interpretable. There are rows for each of the independent variables; in this case, only advertising appears as an independent variable. The *Coefficient* represents the impact of a unit increase in the independent variable upon the response variable, accounting for all other variables in the model. In this case, spending an additional $1000 on advertising per month is estimated to lead to an expected additional $1758 in monthly sales.

Statistical theory (and some logical thinking) tells us that if we took a different random sample of our empirical data (e.g., we redo our survey and, by chance, get different observations), the data points and associated least-squares regression line would look a bit different.[25] The *Standard Error* (Column C) represents the expected (standard) deviation for the values of this coefficient (Column B) when we would repeat the survey. Dividing the Coefficient by this Standard Error gives the t statistic (*t Stat*), which measures how many standard errors away from zero the Coefficient is. So, if we were to hypothesize that the real effect of the X variable on Y is zero (i.e., the Coefficient of the true model is zero, and the only reason our samples provide an estimate different from zero is due to random sampling error), then we can ask how likely it is that a random sample could produce an effect (Coefficient) this large. The answer to that question is the *P-value*. It tells us the probability that chance (i.e., random sampling) alone could produce an estimated effect this strong, under the assumption that the true effect is zero (also often denoted as assuming that the null hypothesis, H_0, is true) If this P-value is sufficiently low, then we would have high confidence to reject this hypothesis and say that we have evidence that the effect of the variable is statistically significant. Alternatively, a high P-value simply means that there is not enough evidence to reject the hypothesis of zero effect—it does not mean that we have evidence that supports a zero effect. A high P-value also does not mean that there is no effect, since the P-value is highly dependent on the sample size.[26]

An alternative to the hypothesis test and P-value is provided by a confidence interval, one indicated by the *Lower 95%* and *Upper 95%* calculations.[27] This provides a 95% confidence interval for the possible values of the Coefficient, based on our sample data. Thus, we may say we are 95% confident that the true impact of a $1000 increase in monthly advertising will lead to an increase in monthly sales that is between $1225 and $2290.[28] But there is nothing magical about the 95% confidence interval or any particular P-value.[29]

The size of the confidence interval is calculated by adding or subtracting an appropriate number of Standard Errors from the estimated Coefficient (assuming a Student-t distribution with seven degrees of freedom, but that goes beyond the current scope). The better the fit and the greater amount of data, the lower this Standard Error will be, and the narrower the resulting 95% confidence interval for the Coefficient will be. A better fit to the data means there is less uncertainty about the true parameters of the regression model.

The P-value and 95% confidence interval are intimately related: they are both based on the Standard Error and estimated Coefficient. If the 95% confidence interval does *not* contain zero, then we are at least 95% confident that the variable is either positively or negatively related to the response variable (in this case, there is a positive association between advertising and sales). In that case, the corresponding P-value will thus be less than 5%. Whenever the 95% confidence interval includes the value zero, then the P-value will be greater than 5%, and we cannot say that there is a "significant" association between the independent and dependent variable.

Use of 5% at the critical P-value for determining statistical significance is fairly standard historical practice, but is increasingly being questioned—and abandoned.[30] In fact, the choice of the threshold P-value is, however, a totally *arbitrary judgment* that the analyst makes. As you can imagine, the choice of a low threshold for the P-value (say 1%) places a high burden of proof for declaring the effect significant. But this means that there is a correspondingly larger risk that we will not find a statistically significant relationship even if one really exists. Conversely, declaring an effect statistically significant at a higher threshold value (e.g., 10%) reduces the risk of failing to acknowledge a relationship in the data, but runs a higher risk (up to 10%) of declaring an effect significant when it really is only the result of random sampling. There is always a trade-off between these two types of potential errors (called Type I and Type II errors[31]), and the conclusion of believing a certain claim should not just be based on a "p < 5%" or "p < 1%" result, but should also take into account all other evidence supporting the claim.

In fact, more and more experts consider the great reliance on P-value inappropriate, and the American Statistical Association (ASA) recently said, among others, that *Scientific conclusions and business or policy decisions should not be based only on whether a p-value passes a specific threshold.*[32] In public health, experts, for example, may conclude that the cost of failing to recognize a significant effect of a potential hazard on human health is greater than the cost of declaring an effect significant when it really is not. Of course, it is easy to imagine cases where the reverse could be true: a particular potential hazard could be of little consequence, but it might be very costly to mitigate the potential impact. Therefore, rather than arbitrarily choosing a P-value threshold, we advise you to carefully think about all of the evidence that you may have, and also think about the relative risks and costs of Type I and Type II errors in the particular issue you are analyzing. Furthermore, no P-value choice needs to be made in a simulation model, as the model takes into account the degree of uncertainty regarding the relationship. We will now illustrate how this can be done.

4.8 Simulation within Regression Models

Returning to our advertising effectiveness case, Figure 4.21 suggests that the regression model may be a good description of how an increasing advertising budget increases sales. The R squared is relatively high, the P-value that the advertising coefficient is zero is very low (P-value = .000006), the relationship makes logical sense, and the residuals appear to be randomly scattered around the regression line.[33] Suppose we wish to forecast the sales level that would result from a monthly advertising level of $7500. We could use the estimated linear regression, Sales = 1.76 + 0.33*Advertising, to derive the Sales forecast. But because of the limited amount of data, and because of chance, there is uncertainty about the true relationship between advertising and sales. Fortunately, the regression analysis output also provides evidence regarding the level of uncertainty in the relationship. Since we have the standard deviations for the intercept (0.33 with standard error = 1.43) and slope (1.76 with standard error = 0.23) and the regression assumptions imply that the true intercept and slope will follow a Student-t distribution (in this case, with seven degrees of freedom) around the sample estimates, we have the information required to simulate the possible intercepts and slopes of the true advertising–sales relationship. There is one complication, however. The uncertainty distributions of the intercept and slope are *not* independent—as the intercept varies, the slope of the least-squares regression line though the original data points will depend on this intercept. If an intercept higher than 0.33 were simulated, we would find that the least-squares regression line through the data points would need to be flatter, resulting in a expected slope lower than 1.76. Conversely, if we simulate a intercept lower than 0.33, then the line will tend to be steeper (i.e., have a higher slope).

To correctly simulate the uncertainty in the relationship, we need to adopt a different procedure than simply simulating the intercept and slopes independently. The method we use and illustrate here is called a *parametric bootstrap*: viewing the current data as one random sample of data points from the underlying process, we will generate additional random samples of new data (a.k.a. bootstrap samples) and estimate the least-squares regression model for each of these bootstrap samples. The resulting simulation then considers full variability inherent in our data.

Regression Model4.xlsx contains this parametric bootstrap model for our advertising-sales data. Generating random samples of sales data relies on the use of the standard error of the residuals around the regression line. This is shown in Figure 4.21 as the Standard Error in the Regression Statistics table (2.09 in this case). It can also be found by using the STEYX function in Excel. Figure 4.22 shows the simulation model.

The first model (in the box in Rows 14 through 20) simply uses the slope and intercept of the least-squares regression line through the original data and then uses the regression standard error to simulate the potential sales that would result from an advertising level = 7.5. The second model (in the bottom box, Rows 22 through 28) estimates the least-squares regression line based on the bootstrap sample in cells F2:F12 (i.e., in this box, the uncertainty around the parameters of the regression line is considered). The bootstrap samples are constructed by simulating the sales level for each advertising level in the original dataset. The standard error of regression is used (in a Normal distribution) to simulate a new sales level, based on the average level from the original regression line.[34] Each bootstrap sample is then used to estimate the least-squares regression intercept (cell C24), slope (cell C25), and standard error of regression (cell C26). These three parameters from fitting a regression to the bootstrap sample are then used to simulate the Sales forecast corresponding to the Advertising level of 7.5 (cell D28).

Figure 4.23 illustrates the original data points and regression line, one possible bootstrap set of data points and corresponding regression line, and one possible value of the predicted Sales for each model that are associated with Advertising = 7.5.

Each simulation (obtained with @Risk by pressing the F9 key after selecting the dice, randomization, button in @Risk) will refresh the picture for a new bootstrap sample and Sales forecast. Running the simulation provides the range of forecast results as shown in Figure 4.24.

The overlaid histograms reveal that the bootstrap forecast is slightly more uncertain than what results from just simulating the Sales level around the static least-squares regression line. This is most easily seen by looking at the standard deviations and the shaded confidence intervals at the top of the graph. The means are extremely close, and with more iterations will converge to be the same. The slightly higher uncertainty in the sales level forecast with the bootstrap

	B	C	D	E	F	G|H	I	J	K	L	M	N	O
2	Data:	advertising	sales		Bootstrap Sample		Advertising	Sales data	BS Samples	Regression	BS Regression	Stnd Forecast	BS Forecast
3		1	2		=$C$16+C3*$C$17+RiskNormal(0,$C$18)		=C3	=D3	=F3	=$C$16+C3*$C$17	=$C$24+C3*$C$25		
4		2	7		=$C$16+C4*$C$17+RiskNormal(0,$C$18)		=C4	=D4	=F4	=$C$16+C4*$C$17	=$C$24+C4*$C$25		
5		3	3		=$C$16+C5*$C$17+RiskNormal(0,$C$18)		=C5	=D5	=F5	=$C$16+C5*$C$17	=$C$24+C5*$C$25		
6		4	7		=$C$16+C6*$C$17+RiskNormal(0,$C$18)		=C6	=D6	=F6	=$C$16+C6*$C$17	=$C$24+C6*$C$25		
7		5	10		=$C$16+C7*$C$17+RiskNormal(0,$C$18)		=C7	=D7	=F7	=$C$16+C7*$C$17	=$C$24+C7*$C$25		
8		6	9		=$C$16+C8*$C$17+RiskNormal(0,$C$18)		=C8	=D8	=F8	=$C$16+C8*$C$17	=$C$24+C8*$C$25		
9		7	13		=$C$16+C9*$C$17+RiskNormal(0,$C$18)		=C9	=D9	=F9	=$C$16+C9*$C$17	=$C$24+C9*$C$25		
10		8	12		=$C$16+C10*$C$17+RiskNormal(0,$C$18)		=C10	=D10	=F10	=$C$16+C10*$C$17	=$C$24+C10*$C$25		
11		9	19		=$C$16+C11*$C$17+RiskNormal(0,$C$18)		=C11	=D11	=F11	=$C$16+C11*$C$17	=$C$24+C11*$C$25		
12		10	18		=$C$16+C12*$C$17+RiskNormal(0,$C$18)		=C12	=D12	=F12	=$C$16+C12*$C$17	=$C$24+C12*$C$25		
13							7.5					=RiskOutput()+D20	=RiskOutput()+D28
14	Forecast without parameter uncertainty:												
15													
16	Intercept	=INTERCEPT(D3:D12,C3:C12)											
17	Slope	=SLOPE(D3:D12,C3:C12)											
18	SE_xy	=STEYX(D3:D12,C3:C12)											
19		x	y					forecast (for total uncertainty):					
20	Forecast	7.5	=C16+C20*C17+RiskNormal(0,C18)					1					=$C$24+I20*$C$25+RiskNormal(0,$C$26)
21								2					=$C$24+I21*$C$25+RiskNormal(0,$C$26)
22	Forecast with parameter uncertainty:							3					=$C$24+I22*$C$25+RiskNormal(0,$C$26)
23								4					=$C$24+I23*$C$25+RiskNormal(0,$C$26)
24	Intercept	=RiskOutput("Intercept")+INTERC						5					=$C$24+I24*$C$25+RiskNormal(0,$C$26)
25	Slope	=RiskOutput("Slope")+SLOPE(F3						6					=$C$24+I25*$C$25+RiskNormal(0,$C$26)
26	SE_xy	=STEYX(F3:F12,C3:C12)						7					=$C$24+I26*$C$25+RiskNormal(0,$C$26)
27		x	y					8					=$C$24+I27*$C$25+RiskNormal(0,$C$26)
28	Forecast	7.5	=C24+C28*C25+RiskNormal(0,C26)					9					=$C$24+I28*$C$25+RiskNormal(0,$C$26)
29								10					=$C$24+I29*$C$25+RiskNormal(0,$C$26)

FIGURE 4.22
Regression Simulation Model

approach is not a surprise, given that it considers an additional level of uncertainty around the parameters of the regression line.

In the present case, the additional amount of uncertainty using the bootstrap procedure is relatively small. This is largely because the linear regression fits this hypothetical data well. The worse the fit, and the less the data, the bigger the difference will be when accounting for this uncertainty. Note that the inherent uncertainty of the parameters of the regression line itself is important to consider—the prediction from the regression line (13.515) account for no uncertainty. Once we recognize that the regression line is only an imperfect (i.e., estimated) description of the relationship between the variables that has inherent uncertainty, the 90% confidence interval for predicted sales ranges from approximately 10 to 17.

4.9 Multiple Linear Regression Models

The methods used in the Advertising–Sales model are readily generalized to models that have a number of independent X variables. We will not provide a complete analysis here, but in

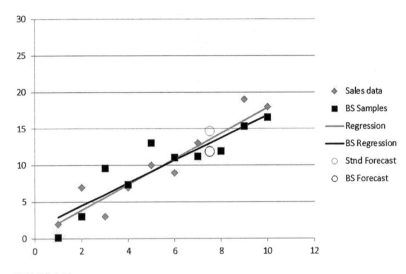

FIGURE 4.23
Bootstrap Sample Regression

order to illustrate the appropriate Excel functions, consider the data (MultipleRegression4.xlsx) and analysis shown in Figure 4.25.

Column A provides hypothetical output data (for example, number of packages sorted per hour at various sorting plants) and data on three independent variables (for example, mechanized sorting lines, number of human sorters, and number of truckers). Use of Regression Analysis from the Data Analysis menu in Excel produces the SUMMARY OUTPUT that is shown in Figure 4.25. Careful inspection of the output shows that the model provides a good fit to the data and that for all three independent variables, assuming their effect on the output is zero, it would be highly unlikely to have obtained the current data by chance (note that Excel will only permit multiple independent variables to be entered into a regression analysis if they are in contiguous columns). The bootstrap procedure illustrated in Section 4.7 can now be used to simulate the output resulting from any particular combination of the three independent variables. However, we cannot use the INTERCEPT, SLOPE, and STEYX functions, as they only apply when there is one single independent variable. Use of the Regression Analysis capability in Excel is also problematic, since each bootstrap sample is different, and we would need to invoke Regression Analysis each time a different sample is generated. This is because Regression Analysis does not automatically recalculate if any of the data changes.

Fortunately, Excel provides the LINEST function that does recalculate as the data changes and produces the required regression model output—however, the resulting outputs from the LINEST function are not automatically formatted as clearly. Figure 4.25 shows the function and its output (in cells F22:I26).

LINEST is an array function: this means that the user must first highlight the required number of cells that the output will occupy (it is five rows of data, and the number of columns equals the number of independent variables, plus one). The first argument in the function asks for the range for the response variable, and the second argument contains the range for the independent variables—with multiple X variables, they must be in a contiguous range. The third argument (TRUE) tells Excel to include an intercept, and the fourth argument (TRUE) tells Excel to produce the full set of regression outputs. *Entering the array formula requires simultaneously selecting of Ctrl + Shift + Enter.* The resulting table is unlabeled, but you will find the relevant parameters that are required for our regression simulation are contained in the LINEST function output. In particular, the top row provides the regression coefficients (note that the coefficients are in reverse order, with the constant being the last value), the second

Modeling Relationships

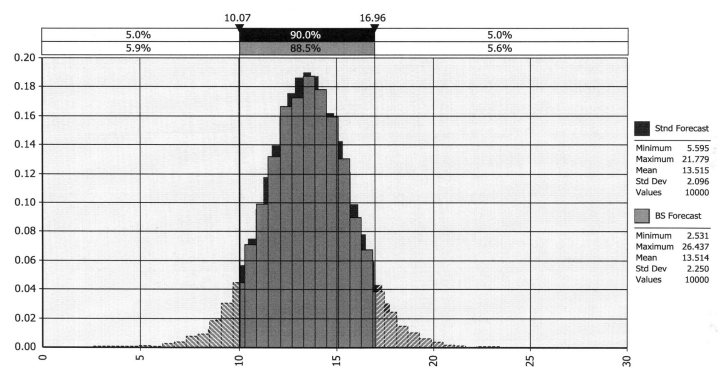

FIGURE 4.24
Regression Simulation Results

row shows the standard errors associated with these coefficients, and the third row gives R squared and the standard error, with additional regression data on the fourth and fifth rows.

We can now use the bootstrap procedure as described in Section 4.7, relying on the standard error and regression coefficients from the LINEST function. The LINEST function will automatically recalculate if any of the input data is changed, which can be used for the bootstrap sample. Thus, LINEST can be useful in simulation models since simulations will usually involve the data changing with each sample. Users are recommended to insert their own text labels

F22 {=LINEST(A2:A31,B2:D31,TRUE,TRUE)}

	A	B	C	D	E	F	G	H	I	J	K	L
1	output	input A	input B	input C		SUMMARY OUTPUT						
2	7668	4	27	16								
3	8204	4	27	17		*Regression Statistics*						
4	11382	6	31	22		Multiple R	0.942359					
5	6666	3	25	13		R Square	0.88804					
6	12328	7	38	16		Adjusted R S	0.875121					
7	17457	8	47	21		Standard Er	1005.561					
8	10016	4	29	16		Observation	30					
9	11581	6	40	26								
10	8628	5	31	14		ANOVA						
11	11774	6	32	23			df	SS	MS	F	ignificance F	
12	13265	5	54	29		Regression	3	2.09E+08	69508445	68.74179	1.72E-12	
13	10918	5	31	26		Residual	26	26289968	1011153			
14	9024	5	33	22		Total	29	2.35E+08				
15	7998	3	25	19								
16	10648	7	34	17			Coefficients	andard Err	t Stat	P-value	Lower 95%	Upper 95%
17	10373	5	34	18		Intercept	739.1736	797.0019	0.927443	0.362227	-899.087	2377.435
18	11003	5	35	24		input A	825.7106	173.2512	4.765973	6.23E-05	469.5876	1181.833
19	14951	8	38	31		input B	94.77352	37.88778	2.501427	0.018996	16.89407	172.653
20	12941	7	39	22		input C	117.9759	38.03784	3.101539	0.004593	39.78796	196.1637
21	13327	6	32	27								
22	7576	3	34	11		117.97585	94.77352	825.7106	739.1736			
23	7013	4	28	6		38.037837	37.88778	173.2512	797.0019			
24	8300	3	26	18		0.8880398	1005.561	#N/A	#N/A			
25	6400	3	19	13		68.741794	26	#N/A	#N/A			
26	14465	8	44	31		208525335	26289968	#N/A	#N/A			
27	6802	3	24	8								
28	9992	4	28	25								
29	6568	3	25	8								
30	7539	3	21	13								
31	12314	8	43	24								

FIGURE 4.25
Multiple Regression Model

next to the LINEST output to remind them of what each number represents.

To implement the bootstrap procedure, the following four steps are required:

1. Use the standard error from the original least-squares model to generate bootstrap samples (use the coefficients of the regression model and add the random term of a Normal distribution with mean 0 and standard deviation = standard error).
2. Use a second LINEST function to estimate the regression model for each bootstrap sample.
3. Use the coefficients and standard error from this LINEST function to estimate the Y value for the combination of X variables you may be interested in.
4. Run a simulation to model the uncertainty inherent in the least-squares regression model.

Within the simulation model, each iteration will generate a new bootstrap sample, re-estimate the regression coefficients, and generate a predicted value of the dependent variable. Thus, running the model for 10,000 iterations provides you with the uncertainty around the value of the dependent variable (e.g., number of packages sorted per hour at various sorting plants).

4.10 The Envelope Method

As we also discussed in earlier chapters,[35] relevant data is not always available, and thus expert opinion is commonly used in Monte Carlo simulation models. Also, the relationship between two (or more) variables in a model can be based on expert opinion, and the envelope method can be a good technique to reflect this in a simulation model. We return to the Facebook stock valuation example from Chapter 3. To illustrate the use of the envelope method, we will focus on one of our experts (Expert #2). Expert #2 provided estimates for revenue growth during the next 5 years of 15%, 22%, and 35% (minimum, most likely, maximum) and estimates for R&D of 20%, 25%, and 30%. The revenue growth and R&D expenses were solicited as independent from each other, but these two variables are likely to be related. Revenue growth is unlikely to be high unless R&D is high—high revenue growth will increasingly attract competitors and Facebook will need to innovate further to sustain high rates of revenue growth.

Therefore, subsequently, we asked Expert #2 to provide information about this potential relationship between revenue growth and R&D spending. For each of the three estimated revenue growth rates (the minimum, the most likely, and the maximum), we solicited a minimum, most likely, and maximum value for R&D that Expert #2 felt would be appropriate. Table 4.2 shows the estimates we obtained.

This table is reproduced in Envelope4.xlsx. The envelope method uses the expert's estimates to take into account the relationship between the two variables. In this case, because we only asked

TABLE 4.2

Expert Opinion for the Envelope Method

For Each Revenue Growth Rate Below, Provide an Estimate for R&D Spending	Minimum R&D	Most Likely R&D	Maximum R&D
15%	8%	11%	13%
22%	20%	25%	30%
35%	23%	31%	35%

Expert B what he/she estimated the R&D expenses to be at three levels of revenue growth rate, we then proceeded to estimate the full relationship using the Excel Data Analysis Regression tool, the LINEST function, or, as we do in this case, by fitting a trendline to this data.[36] We fit a second-order polynomial in order to get a perfect fit through the data in Table 4.2.[37] The fitted trendlines are shown in Figure 4.26.[38]

The envelope method proceeds by simulating the revenue growth rate (using the original Pert distribution obtained from Expert #2), then uses the trendlines to estimate minimum, most likely, and maximum values for R&D (depending on the simulated revenue growth), and finally uses a Pert distribution with these estimated parameters. The model is shown in Figure 4.27.

The trendline equations, obtained from the chart,[39] appear in cells B11:B13 to estimate the parameters for the R&D Pert distribution, based on the simulated revenue growth (cell B14). Running the simulation and using a scatterplot for the simulated revenue growth and R&D values results in Figure 4.28.

Notice how the simulated R&D percentages fall within the ranges shown in Figure 4.26. Not shown are the simulation results for

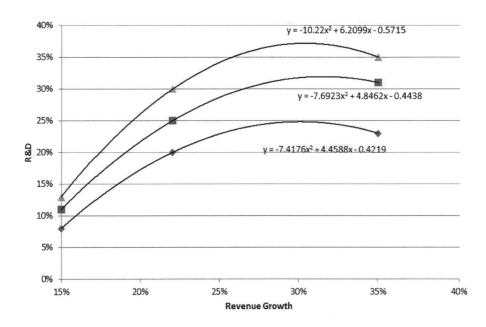

FIGURE 4.26
Trendlines for Expert Opinion

Modeling Relationships

	A	B	C	D
1			R&D%		
2	Revenue Growth	Minimum		Most Likely	Maximum
3	0.15	0.08		0.11	0.13
4	0.22	0.2		0.25	0.3
5	0.35	0.23		0.31	0.35
6					
7	Simulation				
8	Revenue Growth	=RiskOutput("Revenue Growth")+RiskPert(A3,A4,A5)			
9	R&D independent	=RiskOutput("R&D independent")+RiskPert(B4,C4,D4)			
10	trend line estimates				
11	minimum	=-7.4176*B8^2+4.4588*B8-0.4219			
12	most likely	=-7.6923*B8^2+4.8462*B8-0.4438			
13	maximum	=-10.22*B8^2+6.2099*B8-0.5715			
14	R&D fraction with dependency	=RiskOutput("R&D fraction with dependency")+RiskPert(B11,B12,B13)			

FIGURE 4.27
Envelope Method Model

R&D when it is independent of revenue growth, but clearly those values will simply fall within the original range of (20%, 25%, 30%). So, the envelope method provides a means for capturing important dependencies in expert opinion modeling.

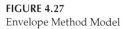

4.11 Summary

We have explored a variety of ways, approaches, and techniques to capture and model relationships among variables in simulation models. As a general rule, it is always best to use the simplest model you can that captures the essential nature of the relationships. Logical statements are powerful and simple to use and can capture many different types of relationships. An example we discussed is where one or multiple variables takes on the value 1 if some condition for another variable is met, such as a drug that is approved or a competitor that launches a product, as this may have an effect on our future sales.

When we have data available, or the number of values is large, rank-order correlation and copulas may capture the interdependence. The rank-order correlation is fast and easy to implement (but not as easy to explain) and permits varying strengths in the relationships between variables to be modeled. More complex copulas should be used when normal (Gaussian) copulas inadequately reflect the data.

If a causal relationship is suspected based on available and relevant data, then regression modeling can be a better approach to modeling the relationship between variables. Regression models can also provide measures of the uncertainty about the relationship, and simulation can incorporate this uncertainty in the relationship between two (or more) variables.

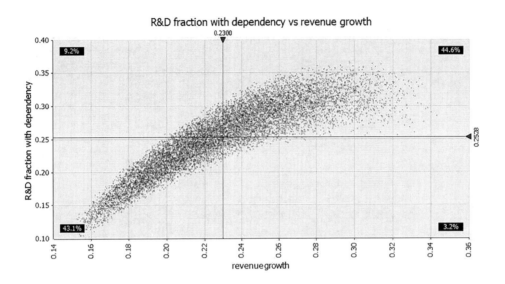

FIGURE 4.28
Envelope Method Simulation

Exercises

Exercise 4.1 Drug Development

Consider a potential pharmaceutical drug for treating type 1 diabetes that is ready to enter phase 1 of research and development for FDA approval. Initially, a 60% chance is assumed that the drug will be effective on any individual patient tested in phase 1. Phase 1 is expected to cost $5 million and will last 1 year. In phase 1, 100 volunteers will be tested and the drug will proceed to phase 2 of the R&D process if more than 50% of the 100 patients are successfully treated (assume the tests on each individual are independent). In phase 2, 250 volunteers will be tested over a 2-year period, and this will cost $10 million. The probability that the drug will be successful in treating type 1 diabetes, if it passed phase 1, is assumed to be 70% for each patient. The drug will proceed to phase 3 of the R&D process if at least 60% of the 250 patients tested are successfully treated for type 1 diabetes.

In phase 3, testing expands to 5000 people for understanding the longer-term effects of the drug and is expected to take 4 years and cost $25 million. If the drug has successfully passed phase 2, the probability it will successfully treat each patient for type 1 diabetes is assumed to be 80%. The drug will be FDA approved in stage 3 if it successfully treats type 1 diabetes in at least 80% of the 5000 people. If approved, the (net) free cash flow is forecasted to be $30 million per year for 15 years for type 1 diabetes if it proves effective for the entire type 1 population.

a. Assuming a cost of capital of 10%, what is the expected net present value (also called eNPV) of this drug development? What is the probability that developing this drug will have a positive NPV?

b. Estimate the NPV and the probability that drug development has a positive NPV, assuming each stage of the development process is successful. Explain the results.

Exercise 4.2 Airfares

Exercise4-2.xlsx contains data for all continental US domestic air routes that averaged at least 10 passengers per day for the first quarter of 2009.[40] Data is provided for the originating and terminating city (and airport code) for each route, along with the average one-way distance, average number of passengers per day, market share of the largest carrier, code for the largest carrier, and the average fare charged on the route.

a. Build a multiple regression model explaining airfares as a function of distance, passengers, and market share of the largest carrier. Interpret the meaning and statistical significance of each coefficient.

b. Suppose, due to consolidation in the airline industry, we expect that the market shares of the largest carriers will increase on average by 10%, and that the average number of passengers per day will increase by 2 (due to better scheduling and flight connections). Provide a point estimate of the impact on the average of the airfares across all routes that would result, using your regression model. Simulate the average fare across all routes without accounting for uncertainty about the regression equation parameters. (*Hint:* Following Section 4.9, use the standard error of the whole regression for this—use the fact that the regression model goes through the means of all the variables, i.e., the average fare as a function of average distance, average passengers, and average market share is a point on the regression line.)

c. Using the uncertainty inherent in the data, perform a parametric bootstrap to simulate the resulting average airfare. Compare the probability that airfares will rise (as a result of consolidation) based on your models with and without parameter uncertainty.

Exercise 4.3 Portfolio Correlation

Exercise4-3.xlsx contains data for the weekly changes in Bitcoin prices (BTC) and the Dow Jones Industrial Average (DJI) over the decade of 2010–2019.

a. Fit a distribution for each index's weekly price change and a parametric copula for the relationship between the two. Use the (lowest) AIC score to choose the distributions and copula that fits best.

b. Simulate the weekly results for each index over 4 weeks.

c. Simulate a $100 investment in BTC, a $100 investment in the DJI, and a portfolio of $50 invested in each. Compare (and explain) the expected returns and probabilities of losing money for each of these three investments over a 4-week period.

Exercise 4.4 Automobile Driving Habits

Exercise4.4.xlsx has data from the 2009 National Transportation Survey for 64,861 gasoline engine automobile annual miles driven and average mpg fuel efficiency.[41]

a. Fit a parametric copula to these variables.
b. Fit an empirical copula to this data.
c. Estimate the fraction of automobiles that get less than 15 miles per gallon and are driven for at least 10,000 miles for each of these copula fits. Compare and discuss your results for a and b with the actual fraction in the data.

Exercise 4.5 Technology Diffusion

The Bass diffusion model is commonly used to forecast adoption of new technologies. It rests on three assumptions: that there is a fixed population of users (who will eventually adopt the new technology), that *innovators* will adopt in proportion to the current number of adopters, and that there are *imitators* whose adoption is proportional to the product of the current adopters and the remaining potential adopters. The origins of the Bass diffusion model lies in epidemiology: infections may spread through a population in a similar fashion (to new technologies or services): innovators are people prone to getting a disease, while imitators catch the disease from the innovators. The number of imitators over time is affected by two factors, working in opposite directions. There are more innovators to "infect" future imitators, but there are fewer potential adopters left to infect. Equation 4.1 reflects the Bass diffusion model, with the assumption that the rate of innovation (p) and the rate of imitation (q) are constant over time:

$$n_t = p(N - N_{t-1}) + qN_{t-1}\left(1 - \frac{N_{t-1}}{N}\right) \quad (4.1)$$

where n_t is the number of new adopters during period t, N_{t-1} is the cumulative adopters are the beginning of period t, and N is the total pool of potential adopters. Equation 4.1 states that the new adopters will come from the innovators (a constant fraction of the remaining non-adopters) and the imitators (a constant proportion of the product of the current adopters and remaining fraction of non-adopters). Equation 4.1 is not in the best form to estimate from actual data, however. Multiple regression can be used to estimate the diffusion curve when it is rewritten in the form of Equation 4.2:

$$n_t = pN + (q-p)N_{t-1} - \left(q/N\right)N_{t-1}^2 \quad (4.2)$$

Equation 4.2 only requires data on total adopters at each point in time, and using multiple regression analysis to estimate $Y = a + bX + cX^2$ will permit estimation of the parameters of the Bass diffusion model.

a. Exercise4-5.xlsx provides annual data for the number of active Facebook users (measured in millions) over its 2004–2018 history (derived from Facebook annual financial filings and other data sources). The total number of users is given, along with its squared value, the number of new users, the lag of the number of users, and its squared value. Estimate the Bass diffusion model by fitting a multiple regression model to this data, and then deriving p and q from the estimated coefficients of that model. Assume that the total number of potential adopters is 3,500,000,000. Typical estimates of the Bass diffusion model for consumer manufactured goods result in estimates of the coefficient of innovation (p) around 3% and estimates for the coefficient of imitation (q) of around 35%. Compare your estimates for Facebook with these values and interpret the differences.

b. Simulate the 2019–2021 end of year total user number, using the regression model and the uncertainty in the data. Provide a 95% confidence interval for your prediction. (*Hint:* Use the bootstrap technique for this, as well as solely using the uncertainty embodied in the standard error of the regression equation.)

c. Now assume that the saturation level of users is uncertain, using a Pert distribution with a minimum of 3 billion, most likely value of 3.5 billion, and maximum value of 5 billion users. Simulate the end of year total users for 2019–2021. Compare your answers in parts (b) and (c).

Exercise 4.6 Global Warming and Economic Development

Carbon emissions (in particular, carbon dioxide) and their relationship with global climate change is a major issue, fraught with uncertainties. Exercise4-6.xlsx has data on per capita carbon emissions, per capita income (GDP, in $US), and total population for 161 countries from 2013. For purposes of this question, assume that there is a need to reduce total carbon emissions by 20%.

a. Using the countries with per capita GDP levels below $5000, what is the total impact on carbon emissions of reducing emissions per person by 30%? What would the impact be if emissions per person are reduced by 30% only in countries with per capita GDP levels above $25,000?

b. Fit distributions to per capita emissions and per capita GDP and for the copula between these. Given that the data is not complete for all countries, these distributions may be thought of as representative of all countries.[42]

c. Provide a probabilistic estimate (use a 90% confidence interval) for the questions in a. Compare your answers to a and b.

Exercise 4.7 Warranties Revisited

Consider Exercise 3.7 concerning tire failures and warranties. There are five sources of uncertainty: tread wearout, sidewall failure, steel shear of tread, punctures, and sale of car. Assume that the correlation between sidewall failure and steel shear of tread is 0.8, between tread wearout and sidewall failures is 0.5, between tread wearout and punctures is 0.6, between punctures and sales is 0.45, and between tread wearout and sale is 0.3.

a. Using @Risk, construct a correlation matrix representing these correlations and check to make sure it is valid.

b. Using a correct correlation matrix, reexamine the design question from Exercise 3.7. Compare your answer with and without any correlations: in particular, is the same design choice the most profitable, and how does the distribution of profits affect the most profitable design, with and without correlations.

Exercise 4.8 Time Use

Exercise4-8.xlsx contains a subset of data from the 2017 American Time Use Survey conducted by the US Bureau of Labor Statistics. This extensive survey contains detailed diary information about how a variety of people spend their time. This excerpted data has only a few variables out of hundreds of activities and only includes working adults between the ages of 25 and 50. These 3473 individuals' daily minutes spent in a variety of activities are provided in the data.

a. For now, ignore the demographic variables—sex and number of children. For the other variables, fit a copula for the minutes per day spent in the various activities. Describe the correlations.

b. Repeat part (a), but separately for men and women. Then repeat it again, but only for men and women with children (note: only one survey is done per household, so the respondents in this survey do not represent parents in the same household).

c. For people with children, fit distributions for the time spent sleeping, cleaning, preparing food, child care, and grocery shopping—we might consider these non-discretionary activities. Fit a copula for these activities and individuals. Estimate the total non-discretionary time spent each day for men and women. What is the probability that this total non-discretionary time is greater for men than for women (or vice versa)?

Notes

1. DiMasi, J. A., Feldman, L., Seckler, A. and Wilson, A. 2010. "Trends in Risks Associated with New Drug Development: Success Rates for Investigational Drugs." *Nature*, 87:3.
2. When n = 1 for the number of trials, the Binomial and Bernoulli distributions are the same thing.
3. Profits are actually much more complex, depending on patient population, disease prevalence, market competition, costs, and so on. The model could be expanded to include these factors, but in this example we use the Pert distribution so that we can focus on modeling the logical relationships between phases of drug development.
4. We have assumed a fixed discount rate—often a difficult input to choose for a profitability analysis over time. Some companies adopt a specific "risk-based hurdle rate" to discount future costs and revenues, while sometimes an opportunity cost of capital (e.g., weighted average costs of capital, WACC) approach is used. If it is not clear what the appropriate discount rate is, then we recommend using a simulation table (or other type of "what-if" analysis) to analyze how the eNPV is affected by the choice of discount rate. We advise *against* making the discount rate a probability distribution, however. It is not random in the same sense as the other uncertainties being modeled.
5. Notice that the probability of success at all stages = 68%*35%*50%*80% = 9.52%, so in the majority of situations (100% − 9.5% = 90.5%), this example drug development investment will be unprofitable.
6. In this case, the asset owned by the CDO may be mortgages or other debt instruments that have an obligation to make payments to the CDO.
7. Each tranche, or slice, of the CDO is made up of a subsection of the underlying assets, and the tranche is then defined based on its level of aggregate risk. Theoretically, a less risky senior tranche may pay less income but is considered to have a lower probability of those assets becoming delinquent or defaulting.
8. In fact, nearly 70% of CDOs were rated AAA, while less than 1% of single-name corporate bonds were rated AAA (SIFMA. 2009. IMF Global Financial Stability Report).
9. Poser, K. A. 2010. *Stalking the Black Swan*. Columbia Business School Publishing, 178–179. This book contains a detailed account of the problems that correlation of housing markets created for CDOs.
10. The CDO4 example can also be approached as a problem of conditional default probabilities, or a modeling problem in which the default probabilities can change over time (instead of using a Gaussian copula to correlate the default events).
11. TRUE in the VLOOKUP function finds the closest value that is greater than the value we are checking—FALSE would look for an exact match.
12. Alternatively, we could have used an IF(OR( )) function for the junior tranche default and the IF(AND( )) function for the senior tranche default. Instead, we check whether the sum of defaults for the two assets is greater than 1 (or equal to 2, respectively) to model the CDO tranche defaults. The results will be identical, but we prefer to avoid combinations of logical statements where possible, due to the ease of making errors when using nested logical functions.

13. We consider more complex forms of correlation in the next section. The Gaussian copula captures a symmetrical correlation.
14. These correlations are Spearman rank correlations—they range from −1 to +1, with 0 indicating no correlation.
15. This data comes from the Case-Shiller home price indices, downloaded from FRED (St. Louis Federal Reserve). Since this is time series data, we should use time series modeling techniques (Chapter 5), but we will just model the monthly changes here, as it is a good illustration of a more complex correlation pattern.
16. Figure 4.11 was generated by using the Data Viewer functionality that is available on the @Risk toolbar.
17. Former Federal Reserve chair, Alan Greenspan, famously cited the local nature of real estate markets prior to the Great Recession, later admitting that this was an error due to the preceding period of constantly rising prices ("Greenspan Admits 'Mistake' That Helped Crisis," Associated Press, 10/23/2008).
18. As a reminder, a small AIC indicates a better fit.
19. Since empirical copulas are not "fitted" to the data, we cannot compare the fit of the ClaytonR and the empirical fit with the use of information criteria.
20. Note that linear regression can also be used to model nonlinear relationships between Y and X. For example, if we define a second X variable to be X^2, then modeling Y to depend on X and X^2 is a linear model (in the two X variables) even though it is a nonlinear model of X alone. A very common approach is to model the logarithm of Y (or another transformation of Y) as a linear function of the logarithm of X, which is a nonlinear function of Y and X (but linear in terms of the logarithms).
21. As mentioned already, readers are urged to refer to any standard statistics book if they require further understanding of the statistical theory behind regression models or detailed information for conducting such analysis. Our focus is on implementing simple regression models in Excel and using Monte Carlo simulation to represent the uncertainty inherent in such models.
22. For example, when the variance of the residuals does vary with the levels of X, weighted regression can often be used to overcome this problem. This technique is covered in any good intermediate statistics text, and can easily be implemented in Excel.
23. Figure 4.20 is obtained by using a scatter chart from the Insert menu in Excel and then adding a trendline (by right-clicking on a point in the chart and selecting "add Trendline."
24. Data Analysis is an Excel add-in that may not be activated when you first install Excel. If it does not appear on the Data menu on the toolbar, go to the Excel start button and click on "Excel options" and then "Add-Ins" and choose "Go" for Manage Excel Add-ins. There is a checkbox for "Analysis ToolPak" that you should check to activate the Add-in.
25. The larger our sample, the less this difference will likely be, which is the reason why, from a statistical perspective, larger samples are typically preferred to smaller ones.
26. There are good resources online that explain the P-value in more detail, including the article "In Brief: The P Value: What It Is and What Does It Tell You?" by Frederick Dorey, https://www.ncbi.nlm.nih.gov/pmc/articles/PMC2895822/.
27. If you produce this display in Excel, you will note that these 95% columns appear twice. That has been the case for every version of Excel. We don't know why these columns are duplicated, and no information is provided by the repeated interval.
28. Strictly speaking, the confidence interval is a property of the *process* of generating the interval. Technically, a 95% confidence interval means that if we take repeated random samples, 95% of the intervals constructed in this way will include the true effect. However, since we only have one sample, we do not know if it is one of those 95% that contain the true effect or one of the 5% that do not. Given that we only have the one sample, we are saying that there is a 95% chance that it is one of the intervals that contain the true effect. This point is subtle, but in statistical terms, the statement in the text is not technically correct.

29. We prefer not to focus on these particular numbers since the usefulness of a regression model really results from the entire range of potential effect sizes and the risks associated with making erroneous conclusions about the relationships being estimated.
30. It is worth your time to read "The ASA's Statement on p-Values: Context, Process, and Purpose" by Wasserstein, R. L. and Lazar, N. A. 2016. *American Statistician*, 70(2): 129–133. The continuing controversy resulted in a petition, with more than 800 signatories, calling for abandoning P-values as decision criteria, "Scientists Rise up against Statistical Significance," *Nature*, March 20, 2019.
31. Even the use of the Type I and Type II error terminology has become controversial. The danger is that it easily leads to dichotomous decision making rules. In an uncertain world, decision making requires examination of risks and their associated probabilities, and careful judgment. This, in large part, is the value of simulation analysis.
32. See https://amstat.tandfonline.com/doi/pdf/10.1080/00031305.2016.1154108.
33. Further diagnostics for the residuals are available and are recommended before settling on a final regression model. In the present case, Figure 4.20 provides a quick visual check that suggests that the assumptions of the linear regression model may be reasonably satisfied with this data.
34. The resulting prediction interval we obtain will be slightly narrower than what you would obtain from a standard statistics package. Linear regressions can be used to analyze data where random observations of X and Y are made together, which is the assumption behind the prediction intervals in most standard statistics packages. However, in the present case, we are not viewing the advertising levels as forming a random sample of advertising—these levels were chosen for particular reasons. Our assumption is that the resulting Sales levels are variable and we are observing but one random sample of what the Sales might be for each level of Advertising. You can read more about this at https://modelassist.epixanalytics.com/display/EA/Linear+regression+parametric+Bootstrap.
35. This section is a bit more advanced and can be skipped without interfering with the continuity of the text.
36. Instead of fitting a trendline, we could, for example, also have interpolated between the three data points with straight lines.
37. Since we are soliciting expert opinion, we are not interested in modeling the parameter uncertainty about the regression fit—so, we eliminate this uncertainty by choosing a quadratic that will fit perfectly through the three data points for the minimum, most likely, and maximum values for R&D spending.
38. The quadratic functions have a counterintuitive property, where there is a negatively sloped portion at the highest revenue growth levels. We will use these here to illustrate the method, but in this case, we would probably choose either a different functional form or a piecewise linear function (i.e., interpolation) to avoid the negatively sloped section.
39. Care must be taken when estimating the trendline equations. The series should be entered as the three points for the minimum estimates, followed by the most likely and maximum estimates. If they are entered as minimum, most likely, and maximum values for each revenue growth rate, a different (and incorrect) trendline equation will result.
40. The data comes from a random sample of 10% of all airline tickets published quarterly by the Office of Aviation Analysis, US Department of Transportation. This data is for the first quarter 2009 Domestic Airline Fares Consumer Report. While more recent data is available, it is more interesting to analyze the 2009 time period, which preceded the latest round of industry consolidation. At that time, the issue of whether airfares would more likely increase or decrease due to impending mergers was a more relevant analysis than it would be currently.
41. This national survey is conducted every 8 years. The 2017 results have been released, but they no longer report fuel efficiency.
42. Not really—this list of countries is not a random sample of countries, so the missing countries would not be expected to follow the same distribution that is represented in this data. For purposes of this question, ignore this fact.

5

Time Series Models

Learning Objectives

- To appreciate the difference between analyzing and modeling distributions versus time series.
- Understand time series forecasting as a data-driven process that assumes a stable underlying pattern in the data.
- Recognize and capture trends, seasonality, and volatility in historical data.
- Learn how to fit time series models to data.
- Learn how to capture the uncertainty in a time series.
- Learn how to simulate time series data for future predictions.
- See how to also use and incorporate (expert) judgment into time series models.
- Learn principles for how to model several interrelated time series.

5.1 The Need for Time Series Analysis: A Tale of Two Series

Imagine we are collecting data where we measure the value of a variable such as the price of grain, sales of a product, or number of birds flying over a certain area, once every interval (day, week, year, etc.) over a period of time. Figure 5.1 shows hypothetical time series data (over 40 periods), along with a histogram (the bars indicating the frequency of values) of the hypothetical data in Figure 5.2.

In addition to a histogram of the data, Figure 5.2 also shows that, if a distribution is fitted to this data, a uniform distribution provides the best fit. Of course, if we are now asked to predict what the next value (in period 41) is, we can't just take a random number from the best-fitting distribution shown in Figure 5.2. In that case, we would predict an average of around 105, the mean of the distribution, with values ranging from 3 to 219. However, this would clearly be a poor prediction. It ignores the seasonal pattern, it ignores the trends, and, in summary, it completely ignores the temporal structure of the data. Indeed, the data in Figure 5.1 was created to show a constant upward

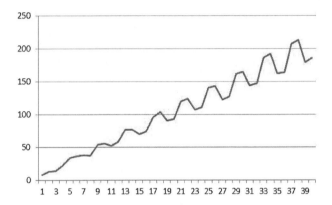

FIGURE 5.1
Hypothetical Time Series Data

trend, with quarterly seasonality that is proportional to each value (and a small random component added).

@Risk treats time series data differently in that the order of the data series is considered. There are three modeling options under "Time Series" on the @Risk toolbar. We can (1) fit a time series model to the data, (2) define a time series model without using data, or (3) batch-fit time series models to multiple series of data (including correlations between them). For now, we will start with fitting a time series model to a single data series that we showed in Figure 5.1.

Figure 5.3 shows the time series dialog that results when trying to fit a time series model to our hypothetical data (see the file called Time Series5.xlsx).

Figure 5.3 shows a warning that the "Mean appears to be non-stationary." This is an important warning and should not be ignored. Stationarity is a property that means that the systematic components of the series have been modeled so that the remaining data can be

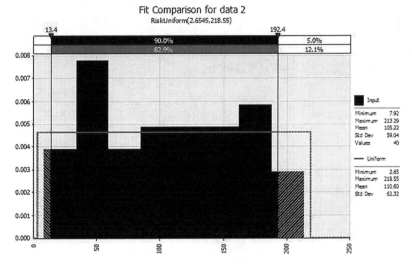

FIGURE 5.2
Distribution of the Time Series

viewed as random variation (the latter is an important requirement when fitting time series models). In fact, if (incorrectly) we ignore the non-stationary nature of the data and proceed with fitting a time series, we find that an ARMA (autoregressive moving average) model provides the best fit, and Figure 5.4 shows the mean forecast along with two confidence intervals for the forecast and one sample simulated forecast path.

Looking at the resulting prediction, it seems indeed that clearly something has gone wrong. @Risk has recognized that there is autocorrelation in the time series but is attempting to model the entire variability as random (without first considering the trend and the seasonality)—hence, the uncertainty ranges are wide and the

Time Series Models

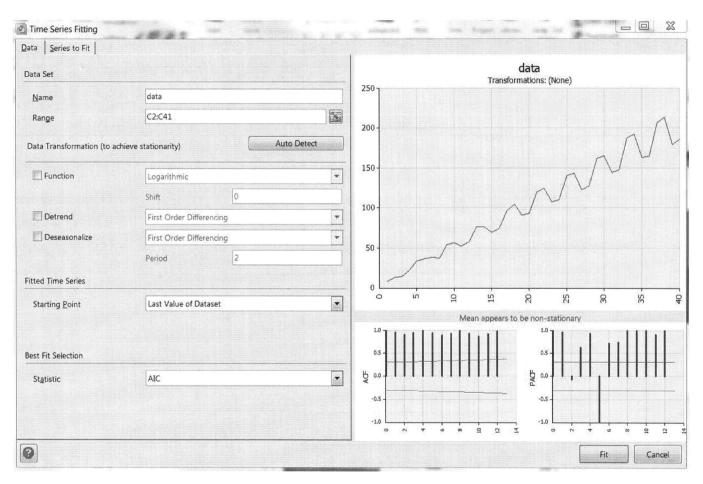

FIGURE 5.3
Time Series Model Dialog

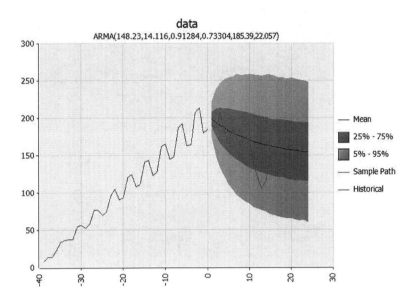

FIGURE 5.4
Fitting a Time Series Model to Non-Stationary Data

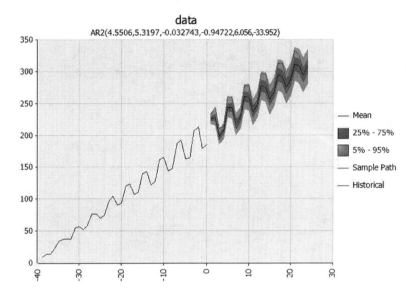

FIGURE 5.5
Forecast for Detrended Data

forecast is moving toward the overall mean of the data series. Before fitting time series models, we need to account for the trend and/or seasonality that may be included and/or evident in the data. This is accomplished with the time series dialog (Figure 5.3) by using the DeTrend and DeSeasonlize options available.

Using the Detrend option, selecting First-Order Differencing, and fitting a time series model provides the forecast series shown in Figure 5.5, and if we also select DeSeasonlize (with "Additive" and period=4, for quarterly data), we get the forecasts shown in Figure 5.6.

First-Order Differencing transforms the data by using the *differences* between successive time series points, rather than the raw data. This often will remove the trend from the data. Additive seasonality is one common way to model seasonality: it adds or subtracts appropriate amounts from the mean forecast depending on the periodicity in the data (an alternative to additive is multiplicative seasonality, which adds or subtracts a percentage from the mean forecast, for example, in each hour in a day or month in a year).

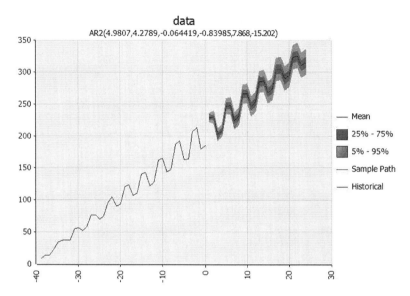

FIGURE 5.6
Forecast with Deseasonalization

There is also an Auto Detect button within @Risk's Time Series Fitting window that can be quite useful—but, as with *any* functionality within @Risk, should be used with care. If we use Auto Detect on this time series, we get the window shown in Figure 5.7.

Notice that @Risk is suggesting including a logarithmic transformation—and provides a warning that both the mean and variance are non-stationary. If you proceed with the Fit (including the logarithmic transformation), you will see that it is not nearly as good as in Figures 5.5 or 5.6. If you deselect the Function transformation in Figure 5.7, then the warning disappears, and you get Figure 5.6. So, Auto Detect can Detrend the data as well as recognizing the additive seasonality with a periodicity of 4. However, its attempted transformation of the data, in this case, was not helpful. Our advice is that you can use Auto Detect, but also make sure to manually adjust the settings to confirm that Auto Detect has successfully resulted in stationarity of the data.

We will now examine several real time series data sets.

5.2 Introductory Case: Air Travel and September 11

The September 11 terrorist attacks had a pronounced effect on air travel. Among the many questions it raised, two are particularly amenable to quantitative analysis and simulation: How much was air travel affected and how long did the effects last? Figure 5.8 shows monthly total revenue passenger mile (RPM) data for US carriers from 1994 through 2006.[1]

This time series seems to clearly exhibit a trend, as well as an annual seasonality pattern. In addition, the data doesn't exactly follow a perfectly repeating pattern, so, unless we have additional data that could explain some of this additional "noise," we would call it (unexplained) randomness. One of such events, though, is the September 11 (2001) event and its impact on air travel. Seasonality is clear from the repeated annual cyclical pattern in the data. In other words, randomness in this context means that variability in the time series can't be explained by factors like the trend, seasonality, and (known) events. An analysis of this air traffic data can give us an

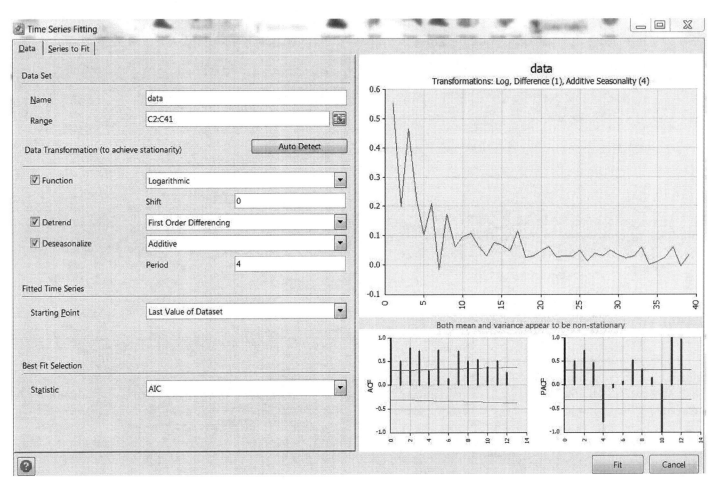

FIGURE 5.7
Auto Detect

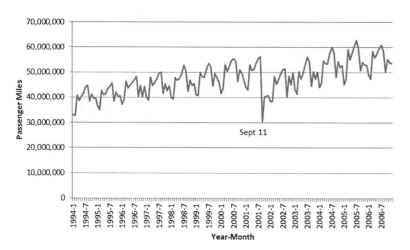

FIGURE 5.8
Air Travel 1994–2006

estimate of the impacts the 9/11 attacks, as well as indicating how long air travel has been impacted.

5.3 Analyzing the Air Traffic Data and 9/11

Our approach to quantifying the effect of the September 11 attacks is to first consider the time series data through August 2001 and use that data to forecast the passenger traffic through 2006 that we would have expected assuming the air traffic volume patterns would have continued as they were through August 2001 if there had not been the September 11 event, and then compare the predicted air traffic volumes with the actual September 2001–December 2006 data.[2] One of the first steps in time series analysis should always be to look at the data patterns over time, and in @Risk, this can be done in the "view the data" screen. Then, choose Time Series–Fit on the @Risk tool bar, and fit a time series model to the data through August 2001. Select Detrend and Additive Seasonality with a period of 12 (using the Auto Detect functionality also results in detrending and deseasonalizing, but with a "period" of 5, which, given the monthly data, we can overwrite unless there is a logical reason for a 5-month repeating pattern), as shown in Figure 5.9.

Air travel5.xlsx contains the data and the fitted time series model. Figure 5.10 shows that the moving average (MA(1)) model fits the air traffic data best, according to the AIC goodness-of-fit measure.

Paste the forecasts into cells D94:D157 (September 2001–December 2006). Run the simulation and view the results by selecting Time Series–Results from the @Risk toolbar, or the Browse Results option with a Summary Trend Chart. Figure 5.11 displays the results.

Our focus is on the impacts of 9/11 on air travel, so we calculate the difference between the actual revenue passenger miles and the mean forecast (Column E), 5th percentile forecast (Column F), and 95th percentile forecast (Column G). Graphing these forecasts produces Figure 5.12.

There are two striking findings in Figure 5.12. First, the mean shortfall is generally around 4–7 million revenue passenger miles per month, a significant impact (on the order of 10% of total RPM). Second, the shortfall only slightly decreases after the first year. Apparently, September 11 caused more than a temporary aberration in air traffic.

There are two important caveats to this conclusion. First, we have not investigated any other possible causes for the observed decline in air

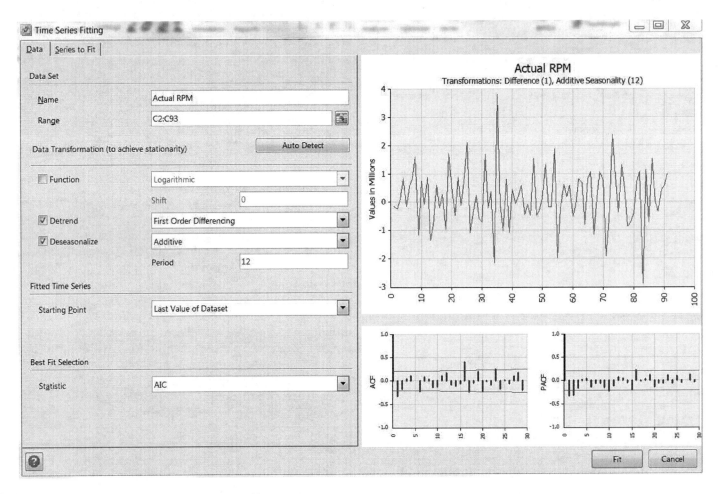

FIGURE 5.9
Fit Time Series for the Air Travel Data

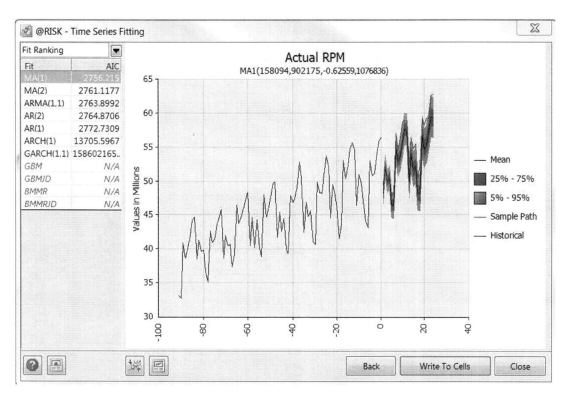

FIGURE 5.10
Time Series Model Fit for the Air Travel Data

traffic, such as a change in the health of the US economy or changes in corporate budgets devoted to air travel (and the potential rise in alternatives to physical air travel using information technology, such as videoconferencing). Second, we have used 8 years of data to predict an additional 5 years—an excessive forecast period for the available data. The implicit assumption that we made by forecasting based on pre–9/11 data is that the historical patterns of seasonality, trend, and variability would remain unchanged over the next 6-year period.

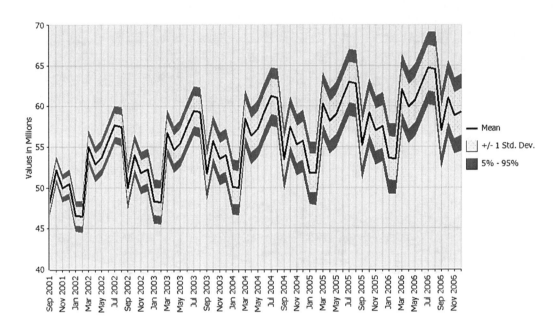

FIGURE 5.11
Air Travel Forecast Trend

5.4 Second Example: Stock Prices

Forecasting stock prices is of considerable interest for many people, particularly to those with equity investments. Stock prices5.xlsx contains 10 years of daily stock price data for Amazon (ticker symbol: AMZN). Figure 5.13 shows the daily closing price and the percentage *change* in Amazon's stock price, calculated as the ratio of natural logarithms of successive daily prices.[3]

The patterns in Figure 5.13 are typical for stock prices: closing prices will typically exhibit an upward trend (at least for publicly traded companies that survive), but the daily changes randomly fluctuate around a mean close to 0. If there were clear and repeatable patterns that could be counted on continuing into the future, we would expect investors to have discovered it, acted on that knowledge (i.e., bought or sold the stock), and as a result, destroyed the patterns. As Figure 5.13 shows, most of the daily fluctuations are within a range of +/−4%, although there are occasional changes on the order of

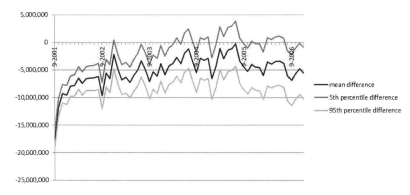

FIGURE 5.12
Actual Estimation of the Monthly Impact of the 9/11 Event on Air Travel

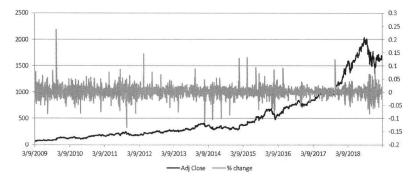

FIGURE 5.13
Ten Years of Daily Stock Prices and Daily Price Changes

6%–8%. Indeed, if we fit a time series model to this data and use Auto Detect in @Risk, the feedback is that the variance appears to be non-stationary, but @Risk is not able to find a suitable transformation. In other words, no trend or seasonality is evident in the daily stock price changes, but there is evidence that the variability changes. This is a more subtle time series pattern that can be modeled.

It is tempting to think that a time series model may not be necessary. So, we will compare fitting a *distribution* to the stock price daily *change* values with fitting a time series model to the data. As we will see, the results will be fairly similar, but not quite the same. Fitting a static distribution to the daily price *change* data reveals the Laplace distribution as having the best fit (according to the AIC). We can fit a time series model directly to the stock prices themselves—Auto Detect tells us that both the mean and standard deviation are not stationary, and uses first-order differencing to provide a stationary mean. In other words, @Risk will fit a time series model to the changes in daily prices. The best-fitting time series model is the GARCH model, as this family of models is designed to capture change in daily volatility, and Figure 5.13 appears to reflect these. We simulate the next 10 days of stock price changes for each of the alternative models and collect the means and a 90% confidence interval for each. Figure 5.14 shows the simulation model.

In order to compare the two approaches, we focus on the changes in daily stock price changes for each. Running the simulation provides the results shown in Figure 5.15.

The results are quite different. The static distribution forecasts a daily price change based on the Laplace fitted distribution that has a wide (and constant) volatility and confidence interval. This is because this model assumes that the variability in stock prices ("volatility") is constant. In contrast, the GARCH model recognizes that there are

	E	F	G
1	Forecast day	Distribution Fit	Time Series Fit (GARCH)
2	1	=RiskLaplace(0.00092952,0.020053,RiskName("Dataset 3"))	=RiskGARCH11(0.61991,0.75637,0.34275,0.76513,-5.1499,0,RiskTsIntegrate(1,1620.800049),RiskTsSync(2,T!$B$2:$B$2519),RiskName("Adj Close"))
3	2	=RiskLaplace(0.00092952,0.020053,RiskName("Dataset 3"))	=RiskGARCH11(0.61991,0.75637,0.34275,0.76513,-5.1499,0,RiskTsIntegrate(1,1620.800049),RiskTsSync(2,T!$B$2:$B$2519),RiskName("Adj Close"))
4	3	=RiskLaplace(0.00092952,0.020053,RiskName("Dataset 3"))	=RiskGARCH11(0.61991,0.75637,0.34275,0.76513,-5.1499,0,RiskTsIntegrate(1,1620.800049),RiskTsSync(2,T!$B$2:$B$2519),RiskName("Adj Close"))
5	4	=RiskLaplace(0.00092952,0.020053,RiskName("Dataset 3"))	=RiskGARCH11(0.61991,0.75637,0.34275,0.76513,-5.1499,0,RiskTsIntegrate(1,1620.800049),RiskTsSync(2,T!$B$2:$B$2519),RiskName("Adj Close"))
6	5	=RiskLaplace(0.00092952,0.020053,RiskName("Dataset 3"))	=RiskGARCH11(0.61991,0.75637,0.34275,0.76513,-5.1499,0,RiskTsIntegrate(1,1620.800049),RiskTsSync(2,T!$B$2:$B$2519),RiskName("Adj Close"))
7	6	=RiskLaplace(0.00092952,0.020053,RiskName("Dataset 3"))	=RiskGARCH11(0.61991,0.75637,0.34275,0.76513,-5.1499,0,RiskTsIntegrate(1,1620.800049),RiskTsSync(2,T!$B$2:$B$2519),RiskName("Adj Close"))
8	7	=RiskLaplace(0.00092952,0.020053,RiskName("Dataset 3"))	=RiskGARCH11(0.61991,0.75637,0.34275,0.76513,-5.1499,0,RiskTsIntegrate(1,1620.800049),RiskTsSync(2,T!$B$2:$B$2519),RiskName("Adj Close"))
9	8	=RiskLaplace(0.00092952,0.020053,RiskName("Dataset 3"))	=RiskGARCH11(0.61991,0.75637,0.34275,0.76513,-5.1499,0,RiskTsIntegrate(1,1620.800049),RiskTsSync(2,T!$B$2:$B$2519),RiskName("Adj Close"))
10	9	=RiskLaplace(0.00092952,0.020053,RiskName("Dataset 3"))	=RiskGARCH11(0.61991,0.75637,0.34275,0.76513,-5.1499,0,RiskTsIntegrate(1,1620.800049),RiskTsSync(2,T!$B$2:$B$2519),RiskName("Adj Close"))
11	10	=RiskLaplace(0.00092952,0.020053,RiskName("Dataset 3"))	=RiskGARCH11(0.61991,0.75637,0.34275,0.76513,-5.1499,0,RiskTsIntegrate(1,1620.800049),RiskTsSync(2,T!$B$2:$B$2519),RiskName("Adj Close"))

	H	I	J	K	L	M	N
1	GARCH change	Static fit mean	GARCH mean	Static 5%	GARCH 5%	Static 95%	GARCH 95%
2	=LN(G2/B2519)	=RiskMean(F2)	=RiskMean(H2)	=RiskPercentile(F2,0.05)	=RiskPercentile(H2,0.05)	=RiskPercentile(F2,0.95	=RiskPercentile(H2,0.95)
3	=LN(G3/G2)	=RiskMean(F3)	=RiskMean(H3)	=RiskPercentile(F3,0.05)	=RiskPercentile(H3,0.05)	=RiskPercentile(F3,0.95	=RiskPercentile(H3,0.95)
4	=LN(G4/G3)	=RiskMean(F4)	=RiskMean(H4)	=RiskPercentile(F4,0.05)	=RiskPercentile(H4,0.05)	=RiskPercentile(F4,0.95	=RiskPercentile(H4,0.95)
5	=LN(G5/G4)	=RiskMean(F5)	=RiskMean(H5)	=RiskPercentile(F5,0.05)	=RiskPercentile(H5,0.05)	=RiskPercentile(F5,0.95	=RiskPercentile(H5,0.95)
6	=LN(G6/G5)	=RiskMean(F6)	=RiskMean(H6)	=RiskPercentile(F6,0.05)	=RiskPercentile(H6,0.05)	=RiskPercentile(F6,0.95	=RiskPercentile(H6,0.95)
7	=LN(G7/G6)	=RiskMean(F7)	=RiskMean(H7)	=RiskPercentile(F7,0.05)	=RiskPercentile(H7,0.05)	=RiskPercentile(F7,0.95	=RiskPercentile(H7,0.95)
8	=LN(G8/G7)	=RiskMean(F8)	=RiskMean(H8)	=RiskPercentile(F8,0.05)	=RiskPercentile(H8,0.05)	=RiskPercentile(F8,0.95	=RiskPercentile(H8,0.95)
9	=LN(G9/G8)	=RiskMean(F9)	=RiskMean(H9)	=RiskPercentile(F9,0.05)	=RiskPercentile(H9,0.05)	=RiskPercentile(F9,0.95	=RiskPercentile(H9,0.95)
10	=LN(G10/G9)	=RiskMean(F10)	=RiskMean(H10)	=RiskPercentile(F10,0.05)	=RiskPercentile(H10,0.05	=RiskPercentile(F10,0.9	=RiskPercentile(H10,0.95)
11	=LN(G11/G10)	=RiskMean(F11)	=RiskMean(H11)	=RiskPercentile(F11,0.05)	=RiskPercentile(H11,0.05	=RiskPercentile(F11,0.9	=RiskPercentile(H11,0.95)

FIGURE 5.14
Stock Price Simulation Model

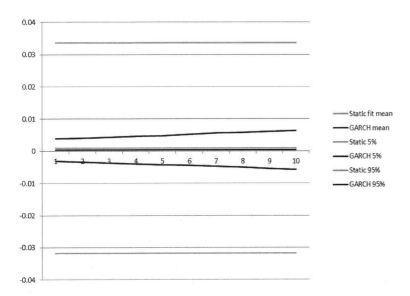

FIGURE 5.15
Comparison of Static and Time Series Models for the Stock Price Data

periods of lower and higher volatility and that the recent period is one of relatively low volatility. Hence, the forecast confidence interval of daily change over the next 10 days is much narrower. The GARCH model does exhibit increasing uncertainty each day. If we extend the forecast period far enough, the GARCH model will eventually be more variable than the static distribution forecast, since the likelihood of a high-volatility period will be increasing.

GARCH type models are designed to recognize and consider changes in volatility over time, and if you think this is an accurate reflection of the stock price behavior, it could be a good forecasting model to use. So, while the level of a stock price is highly uncertain, there are time series methods that can consider (historical and) future changes in the degree of volatility.

5.5 Types of Time Series Models

A wide variety of stochastic time series models have been developed, but most can be grouped into a few major types. These are:

Geometric Brownian motion (GBM) models: these models assume that the natural log of the *changes* (also called the log returns) in a variable between consecutive periods is independent and follows a Normal distribution. These GBM models are sometimes referred to as "random walk models" (even though the GBM is but one type of a random walk model), since each step in the time series is calculated as a random change from the prior value. The change that occurs in one period is not affected by, and does not affect, any previous or future changes. Also, the Black-Scholes formula to calculate the value of call options assumes that the price series follows a GBM model.

Autoregressive moving average (ARMA) models: these assume that the expected (natural log) change in a variable is proportional to its change in the previous recent periods (autoregression), and that the expected change of the variable differs from the long-run mean by a factor proportional to its recent variation (moving average) from its long-run mean. In other words, within these time series methods, the change that occurs in one time period is *affected* by what happened in the last one or two (or more) periods.[4]

Autoregressive conditional heteroskedasticity (ARCH and GARCH) models: within these time series, the volatility of the time series is a function of the previous deviations of the variable from its long-run mean. In other words, in these time series, the volatility can actually change over time.

Each general class of time series model also has several subset models, for example, based on whether there is assumed to be one-period or two-period dependence in the series. Some accommodate seasonal patterns, while others can also take into account large random jumps in the series, also called jump diffusion.

Expert opinion-based models: in practice, most time series forecasts are not purely based on historical data, but either on expert opinion or a combination of historical data and expert opinion.

The next section provides an example of this hybrid approach.

5.6 Third Example: Soybean Prices

Soybeans5.xlsx contains monthly soybean prices from 1988 through 2018. In 2018, a trade war was developing between China and the United States, and China is a major export market for US soybeans. In July 2018, China placed a 25% tariff on US soybeans, and by November, US soybean exports to China had virtually disappeared. Suppose we were constructing a forecast of US soybean export prices in June 2018. Figure 5.16 shows the monthly soybean export prices through June 2018.

Constructing a time series forecast from this data for the remainder of 2018 is likely to be predicting price indices of around 200, since this is where prices were during the spring of 2018 (the index is set at 100 in the base year 2000). A hybrid forecasting model could use the time series forecast from the Figure 5.16 data as a baseline and then incorporate what the potential impacts of a trade war on the prices might look like. The first step is to fit a time series model to the data through June 2018. Auto Detect recommends first-order differencing and fits an ARCH model. We forecast 6 months of prices—this constitutes our baseline forecast.

The hybrid approach now also considers what impacts of a trade war might look like. In the short run, the elasticity of supply for soybeans is virtually perfectly inelastic: that is, farmers cannot adjust their production since their crop has already been planted. The potential demand reduction with a trade war is 16% (China soybean imports ranged from 14% to 18% of total US soybean exports over the 2009–2018 period, so we will use a Pert distribution with 16% as the most likely value). An estimate for the elasticity of demand for US soybean exports is −0.9 (with a standard error of around 0.18).[5] The potential price change is then given by the percentage

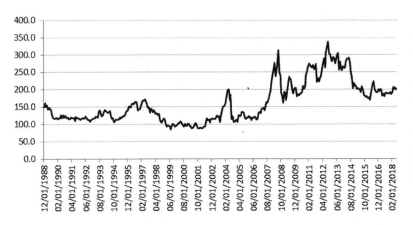

FIGURE 5.16
Soybean Export Prices 1988–2018

Figure 5.18 shows the results running the simulation and comparing the time series forecast model and the hybrid model with the actual price data for the last 6 months of 2018.

Clearly, the baseline forecast of prices does not take into account the effect of the 25% tariffs on soybean imports from the United States to China. However, the mean forecasts from the hybrid model are quite close the actual prices, and the actual prices are well within the 90% confidence interval of the hybrid forecast. The hybrid approach uses the time series data as a baseline, but then attempts to model potential circumstances that might cause a change from the continuation of historical data patterns.

change in demand divided by the price elasticity of demand.[6] Our simulation model simultaneously simulates the drop in demand and the uncertain demand elasticity to calculate the effect on the price forecast. The simulation section of our model is shown in Figure 5.17.

5.7 Fourth Example: Home Prices and Multivariate Time Series

In Chapter 4, we looked at the importance of modeling relationships between uncertain variables that are included in a simulation

	C	D	E	F	G	H
364	% demand interruption	=RiskPert(-0.18,-0.16,-0.14)				
365	demand elasticity	=RiskNormal(-0.9,0.18)				
366	% price change	=D364/D365				
367	Forecast - no change	Forecast - trade war	mean hybrid forecast	0.05	0.95	mean no change forecast
368	=RiskARCH1(0.13927,66.587,0.76939,-5.3,RiskTsIntegrate(1,199.7),RiskTsSync(2,Sheet1!$B$8:$B$362),RiskName("value"))	=C368*(1-$D$366)	=RiskMean(D368)	=RiskPercentile($D368,F$367)	=RiskPercentile($D368,G$367)	=RiskMean(C368)
369	=RiskARCH1(0.13927,66.587,0.76939,-5.3,RiskTsIntegrate(1,199.7),RiskTsSync(2,Sheet1!$B$8:$B$362),RiskName("value"))	=C369*(1-$D$366)	=RiskMean(D369)	=RiskPercentile($D369,F$367)	=RiskPercentile($D369,G$367)	=RiskMean(C369)
370	=RiskARCH1(0.13927,66.587,0.76939,-5.3,RiskTsIntegrate(1,199.7),RiskTsSync(2,Sheet1!$B$8:$B$362),RiskName("value"))	=C370*(1-$D$366)	=RiskMean(D370)	=RiskPercentile($D370,F$367)	=RiskPercentile($D370,G$367)	=RiskMean(C370)
371	=RiskARCH1(0.13927,66.587,0.76939,-5.3,RiskTsIntegrate(1,199.7),RiskTsSync(2,Sheet1!$B$8:$B$362),RiskName("value"))	=C371*(1-$D$366)	=RiskMean(D371)	=RiskPercentile($D371,F$367)	=RiskPercentile($D371,G$367)	=RiskMean(C371)
372	=RiskARCH1(0.13927,66.587,0.76939,-5.3,RiskTsIntegrate(1,199.7),RiskTsSync(2,Sheet1!$B$8:$B$362),RiskName("value"))	=C372*(1-$D$366)	=RiskMean(D372)	=RiskPercentile($D372,F$367)	=RiskPercentile($D372,G$367)	=RiskMean(C372)
373	=RiskARCH1(0.13927,66.587,0.76939,-5.3,RiskTsIntegrate(1,199.7),RiskTsSync(2,Sheet1!$B$8:$B$362),RiskName("value"))	=C373*(1-$D$366)	=RiskMean(D373)	=RiskPercentile($D373,F$367)	=RiskPercentile($D373,G$367)	=RiskMean(C373)

FIGURE 5.17
Soybean Export Price Forecast Model

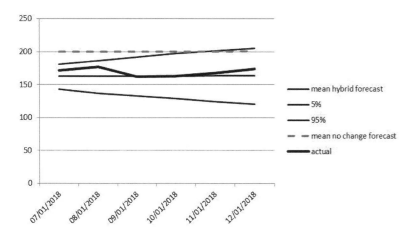

FIGURE 5.18
Soybean Price Forecast Results

model. In similar fashion, we must be aware that there can also be a relationship between random variables being modeled by time series model. Multiple time series, while fluctuating over time (due to randomness, trends, and seasonality), may also be related. For example, we now know that prices of geographic housing markets can be related to one another. These relationship do not need to be causal: it makes no sense to think of housing price changes in Chicago causing house prices to change in Boston (or vice versa). Home prices5a.xlsx has 30 years of monthly data for housing prices in five US cities (San Francisco, Boston, New York, Chicago, and Los Angeles).[7] These are indices of home prices but are not adjusted for inflation. We will examine the monthly price changes after adjusting for inflation—specifically, we calculate the percentage change in monthly home prices and subtract the monthly change in the consumer price index (CPI). We wish to model these related time series, and for that we use a *multivariate time series model* in order to not only include the time series patterns, but also capture the relationship between home prices in the five cities.

Figure 5.19 shows the data. While the individual series are hard to discern, the patterns of the different cities seem similar, and there appears to be some correlation between the price changes in the five cities, but they are far from perfectly correlated.

@Risk has the capability to simultaneously model these time series and estimate and include the correlation between them. This functionality can be found under Time Series–Batch Fit on the toolbar. Selecting the five time series and Auto Detect–Fit creates a new workbook with worksheets for each city and a Summary worksheet (Home prices5b.xlsx). Examining the individual city sheets reveals that @Risk has detrended and deseasonalized the data for each city (using the Auto Detect functionality) and shows forecasts for a number of time series models fit to that city's data. The Summary sheet shows the best-fitting time series model for each city, 2 years of monthly price change forecasts, and a correlation matrix. Clicking on any of the forecast cells will show the time series model with the respective transformations, and includes the correlation matrix to capture the relationships among the different cities.

We will simulate 1 year of monthly price changes for a $100 investment in homes in each city. We also calculate the probability that home prices will fall over the year for each city, along with the number of cities that will simultaneously experience home price decreases. The model section of our spreadsheet is shown in Figure 5.20.

HomePrices5b.xlsx also contains a worksheet with the identical model but with the correlations removed. This will allow us to compare results with and without correlation between the city home prices. Running the simulation reveals that, based on the time series models fitted to the 30 years of real estate pricing data, the probability of prices decreasing ranges from 45% (Chicago) to

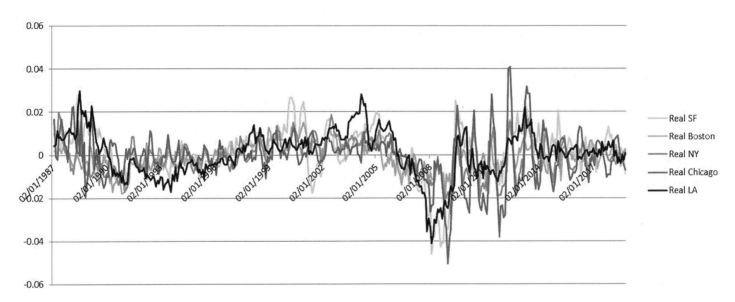

FIGURE 5.19
Monthly Home Price Changes in Five Cities

74% (San Francisco). The results without correlation do not change materially (the correlations among these five cities are relatively small). The number of cities that simultaneously will experience decreases is shown in Figure 5.21.

With correlations, the expected number of cities with price decreases is 2.87 and the standard deviation of the number is 1.118 (without correlations, is it 2.86 and 1.09, respectively). While the results numerically are almost indistinguishable, the direction of change makes sense—correlation makes extreme outcomes (zero or five cities experiencing price decreases) more likely (the probability that all five cities will have price decreases is around 8% with correlation and 6% without). Similarly, there is around a 2% chance that no cities will have a price decrease when correlation is included, but only 1% when it is not. The most likely outcome is three of the five cities having decreases (probability around 32% with correlation and 35% without).

The power of multivariate time series models is that they can easily estimate both the time structure and correlation between these series. This would be very important, for example, for an investor who has real estate holdings in several states. The positive correlation between price changes in the five cities makes the portfolio riskier, while negative correlations (such as between Los Angeles and Chicago) can mitigate overall portfolio risk. In other words, ignoring the correlation could cause risks to be estimated incorrectly.

	B	C	D	E	F	G
43	Month	100	100	100	100	100
44	1	=C43*(1+C11)	=D43*(1+D11)	=E43*(1+E11)	=F43*(1+F11)	=G43*(1+G11)
45	2	=C44*(1+C12)	=D44*(1+D12)	=E44*(1+E12)	=F44*(1+F12)	=G44*(1+G12)
46	3	=C45*(1+C13)	=D45*(1+D13)	=E45*(1+E13)	=F45*(1+F13)	=G45*(1+G13)
47	4	=C46*(1+C14)	=D46*(1+D14)	=E46*(1+E14)	=F46*(1+F14)	=G46*(1+G14)
48	5	=C47*(1+C15)	=D47*(1+D15)	=E47*(1+E15)	=F47*(1+F15)	=G47*(1+G15)
49	6	=C48*(1+C16)	=D48*(1+D16)	=E48*(1+E16)	=F48*(1+F16)	=G48*(1+G16)
50	7	=C49*(1+C17)	=D49*(1+D17)	=E49*(1+E17)	=F49*(1+F17)	=G49*(1+G17)
51	8	=C50*(1+C18)	=D50*(1+D18)	=E50*(1+E18)	=F50*(1+F18)	=G50*(1+G18)
52	9	=C51*(1+C19)	=D51*(1+D19)	=E51*(1+E19)	=F51*(1+F19)	=G51*(1+G19)
53	10	=C52*(1+C20)	=D52*(1+D20)	=E52*(1+E20)	=F52*(1+F20)	=G52*(1+G20)
54	11	=C53*(1+C21)	=D53*(1+D21)	=E53*(1+E21)	=F53*(1+F21)	=G53*(1+G21)
55	12	=C54*(1+C22)	=D54*(1+D22)	=E54*(1+E22)	=F54*(1+F22)	=G54*(1+G22)
56						
57	Decrease?	=IF(C55<C43,1,0)	=IF(D55<D43,1,0)	=IF(E55<E43,1,0)	=IF(F55<F43,1,0)	=IF(G55<G43,1,0)
58	Probability of decrease	=RiskMean(C57)	=RiskMean(D57)	=RiskMean(E57)	=RiskMean(F57)	=RiskMean(G57)
59	Number of decreases	=RiskOutput("Number of price decreases (with correlation)")+COUNTIF(C57:G57,"=1")				

FIGURE 5.20
Five City Home Price Simulation

Exercises

Exercise 5.1 Drug Overdoses

Exercise5-1.xlsx contains monthly data on overdose deaths for a number of drugs.[8]

a. Build an appropriate time series model for total opioid overdose deaths for the 12 months starting in August 2018. Discuss whether there is a trend and/or seasonality and any adjustments you made as a result.

b. Fit multivariate time series models to all drug categories. Compare your results with question a for opioids.

c. Assuming current trends and patterns will continue, estimate the probability that overdose deaths will be lower after 12 months (than in July 2018) for each drug.

Exercise 5.2 Airline Finances

Exercise5-2.xlsx has data since 1995 on airline passenger revenues per available seat mile (PRASM).[9]

a. Adjust the data for inflation (use the CPI data in the CPI worksheet).[9]

b. Build a time series model for the overall (all sectors) PRASM and forecast the 2018 and 2019 expected PRASM

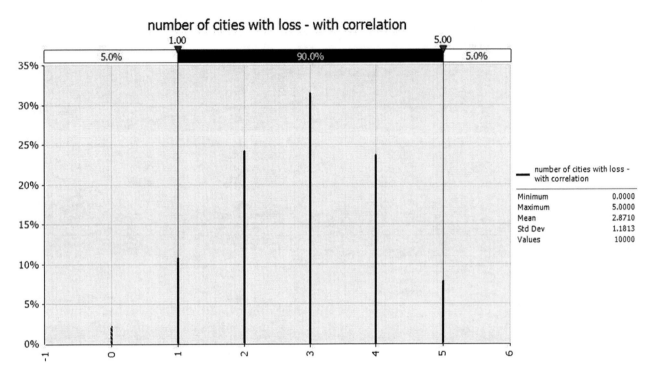

FIGURE 5.21
Simulation Results for Number of City Home Price Decreases

and provide a 90% confidence interval for your forecasts in 2018 and 2019.

c. Repeat part (b) for the three subsectors (network carriers, low-cost carriers, other carriers). Does there appear to be correlation among the three sectors?

d. Within the network airline subsector there have been large company mergers since 1995: Continental with United (2010), Delta with Northwest (2009), and American with both US Airways (2013) and America West (ignore this one for this question). For the first two mergers, use time series models to estimate whether the merger has led to a change in the PRASM (for this exercise, you can assume that any changes in the PRASM patterns are caused by the mergers).

Exercise 5.3 Strangle Options

A *long strangle* is a financial strategy aimed to capitalize on volatility in stock prices. It consists of buying a call option and selling a put option on the same security that expire on the same date. The call option gives the buyer the right to purchase an underlying asset for a particular price (called the strike price) at a particular point in time. A put option gives the seller the right to sell the asset at a specified price at a particular point in time. Thus, if an investor buys a call option at a higher strike price than they sell a put option, they will profit if the stock price is sufficiently volatile. Figure 5.22 illustrates the long strangle.

When the stock price rises above the call option's strike price ("call" in Figure 5.22), then the option will be exercised and the gain will equal the difference between the actual stock price and the call strike price. When the stock price falls below the put option's stock price ("put" in Figure 5.22), then the option will be exercised and the gain will be the difference between the put strike price and the actual stock price. If the price stays between the two strike prices, then the options will not be exercised and no gain will be realized—in fact, a loss will be made, since purchasing these options costs something. So, the long strangle is a gamble that a stock price will deviate substantially from its current value.

Exercise5-3.xlsx has 3 years of daily closing prices for Bitcoins (BTC). Recent experience with Bitcoin prices might suggest that a long strangle would be an appropriate strategy.

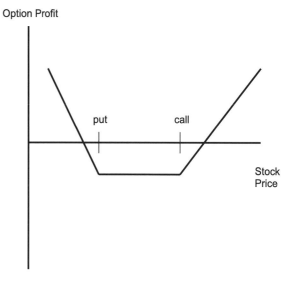

FIGURE 5.22
Strangle Option

a. Build a time series model to predict the daily Bitcoin price for the next 100 days of trading in 2019.[10]

b. Consider a strangle option with a call strike price 10% higher than the March 10, 2019, price and a put strike price 10% lower than the March 10, 2019, price. What is the expected value of this option? In other words, how much should you be willing to pay to engage in this strangle option? Provide a 90% confidence interval for this value, assuming the option is a 25-day, 50-day, 75-day, or 100-day option.

c. Look up the actual Bitcoin price 100 days beyond the end of the data provided. If you had paid the expected value of this strangle option, how would you have fared?

Exercise 5.4 E-Commerce and Retail Sales

Exercise5-4.xlsx has monthly retail sales data for E-commerce and non-E-commerce sales. Clearly, E-commerce makes up an increasing portion of retail sales activity. Ignore inflation until part (d).

a. Construct time series models for the two series independently of each other. Produce a mean forecast, and 90% confidence interval, for the next 12 months of sales for each series.
b. Construct a time series model, accounting for the correlation between the two series, and re-estimate your answers to part (a). Compare your answers and interpret the differences in light of the correlation between the two series.
c. Repeat part (b), but produce time series forecasts for the ratio of E-commerce to non-E-commerce retail sales.
d. Repeat parts (a) through (c) using inflation-adjusted data (i.e., real dollars rather than nominal dollars) and compare your answers (see Exercise 5.2 about adjusting for inflation).

Exercise 5.5 Moore's Law

Moore's Law refers to the empirical observation (by Gordon Moore, cofounder of Intel) that the number of transistors on a chip doubles every 2 years. Exercise5-5.xlsx contains data on CPUs and GPUs for the surface areas of chips and the number of transistors that fit on them.

a. The data will require some cleaning. Within @Risk, time series models require that the time variable be evenly spaced. This means that the missing years will need to be inserted—use a linear interpolation between the available years. Another modification you will need to make is to combine multiple observations for some years into a single observation—use a simple arithmetic average to make this adjustment (the multiple observations derive from data from several manufacturers that may use different processes).
b. Fit time series models to the CPU and GPU data separately and provide a 5-year forecast for the transistors/mm^2 (mean and 90% confidence interval). Then do the same for a multivariate time series (that accounts for correlation between the CPU and GPU). Compare the results.
c. The end of Moore's law has been predicted for a long time. Use your data through the year 2010 to forecast the 2011–2018 data and discuss whether Moore's law appears to be reaching its end based on the comparison of the forecast and the 2011–2018 actuals.

Exercise 5.6 Peak Oil

Predictions that world oil production will peak and decline have been a recurring theme. The theory was originally described by M. King Hubbert, who had predicted that US oil production (from conventional fields) would peak around the year 1970. Exercise5-6.xlsx contains monthly data on oil production from 1973–2018.[11]

a. Produce a time series model for world oil production and estimate world oil production for the next 10 years. Provide a probability each year that total production declines relative to production in 2018. What are the key assumptions in your analysis? Do you think these assumptions are valid?

b. Hubbert's theory really applies to individual oil fields. Your data does not provide production from individual fields, but it does have data broken down by world region. Construct a multivariate time series model for the major producing regions of the world. Estimate the probability that production in each region will decline over the next 10 years. What are the key assumptions in your analysis?

Exercise 5.7 New Car Sales

Exercise5-7.xlsx has monthly data on total new car registrations in the EU15 countries from 2002 through 2018 and US car sales data from 1976 through 2018.[12] You wish to forecast the number of sales for the last month of 2018 and first 6 months of 2019.

a. Visually investigate the time series and comment on trends, seasonality, volatility, and randomness.

b. Fit a time series model to the data for each region separately and explain your choice of model. Fit a multivariate time series model and comment on the correlation (or lack thereof). What are your assumptions?

c. Simulate the seven forecast months of sales in the EU and United States. Then, find actual data and compare your forecast with the actual data—in terms of the mean forecast and confidence intervals for that forecast.

Exercise 5.8 The Olympic Swimsuit Controversy

On Feb. 13, 2008, the LZR Racer swimsuit was introduced that was claimed to reduce race times by 1%–2%. By August 14, 2008, 62 world swimming records had been broken. Exercise5-8.xlsx provides winning times for the Olympic Gold medals in two sports: the men's 100 m freestyle swim, and the men's 100 m track and field race.

a. Fit appropriate time series models to the two series, using the data from 1896–2004. Forecast the winning 2008 times and provide a probabilistic assessment.

b. Based on this data alone, do you think the recent banning of these swimsuits in Olympic competition is justified? Discuss the main assumptions you made and how you might possibly refine your analysis.

c. Include the data from 2012 and 2016 and assess whether the 2008 results seem out of line with the other Olympic results.

Notes

1. Data from the US Department of Transportation, Research and Innovation Technology Administration, Bureau of Transportation Statistics, http://www.transtats.bts.gov/, accessed on December 31, 2010. Later comparable data is not available because it now includes regional jets (<60 passengers), whereas the earlier data did not.

2. We generally do not recommend using @Risk's time series forecasting for predictions that are more than 25%–30% of the available historical data—in this case, around 2 years. Since we have the data through 2006, we will extend our forecast that far, but with the caveat that forecasting this far beyond the 8 years of historical data is typically not advised.

3. This is calculated as $(\ln(\text{price}_{\text{day}_t}) - \ln(\text{price}_{\text{day}(t-1)})$, which equals $\ln(\text{price}_{\text{day}_t}/\text{price}_{\text{day}(t-1)})$, which is the continuously compounded change, and therefore almost the same as the discrete calculation of a percentage change, $(\text{price}_{\text{day}_t} - \text{price}_{\text{day}(t-1)})/\text{price}_{\text{day}(t-1)}$, for small changes.

4. ARMA models generally can consider what happened more than two time steps ago, but within @Risk, they are limited to looking back up to two steps.
5. Adapted from Reimer, J. J., Zheng, X. and Gehlhar, M. J. 2012. "Export Demand Elasticity Estimation for Major US Crops." *Journal of Agricultural and Applied Economics*, 44(4):501–515.
6. The price elasticity of demand is defined as the percentage change in the demand divided by a percentage change in price. In this case, we have the percentage change in demand and an estimate of the price elasticity and wish to find the resulting percentage change in price.
7. The data comes from the "S&P/Case-Shiller Home Price Indices" downloaded from https://fred.stlouisfed.org.
8. National Center for Health Statistics, Provisional Drug Overdose Death Counts, https://www.cdc.gov/nchs/nvss/vsrr/drug-overdose-data.htm, accessed on March 11, 2019.
9. To adjust for inflation, you can recalculate the monetary series to any common year. For example, to express everything in 2018 dollars, multiply each of the nominal dollar figures by ($CPI_{2018}/CPI_{year t}$) for each year t. You can replace 2018 with any other year to express the data in that other year's dollars. We recommend using the most recent year because the value of that year's purchasing power is most intuitive.
10. If you fit a time series to the daily price changes rather than the prices themselves, then you can calculate the prices from the changes using the equation to calculate the predicted stock prices by using the fact that each day's stock price = previous day's price x $e^{\%\ daily\ change}$. This comes from the fact that the % change is calculated as ln ($price_t$/$price_{t-1}$) so that $price_t = e^{\%change}\ price_{t-1}$. Use the EXP function in Excel.
11. Data is from the US Energy Information Administration.
12. The US data source is found on the spreadsheet. The EU data comes from the European Automobile Manufacturers Association, and can be found at https://www.acea.be/statistics/tag/category/passenger-cars-registrations.

6
Additional Useful Techniques

Learning Objectives

- Learn how to use advanced sensitivity analysis to understand how distributions and parameters affect outcomes.
- Use stress testing to explore granularity of outcomes, particularly in the tails of outcome distributions.
- Understand how to use empirical distributions to model uncertainty when parametric distributions do not adequately represent uncertainty.
- Learn how to use Combined distributions in order to conduct frequency-severity modeling.
- Use Combined distributions to correctly model insurance problems.
- Appreciate the applicability of frequency-severity modeling to a variety of decision problems.
- See how Bayesian modeling allows the analysis of uncertainty that will be updated as new information becomes available.

6.1 Advanced Sensitivity Analysis

We have already seen how tornado charts can be used to understand which uncertainties in our models are the greatest risk drivers of the model outputs and how various uncertain inputs can affect those model outputs. Such information helps identify which are the key uncertainties and can therefore guide avenues that might be used to most effectively reduce uncertainty. By concentrating resources in an area where the uncertainty matters the most for the outcome, we can most efficiently reduce the variability in our outputs. For example, we saw that the uncertainty in customer sensitivity has a much larger impact on MoviePass profitability (to use the lead example in Chapters 1 and 2) than the uncertainty in increasing movie attendance, so market research may be more effective in reducing risk if targeted at understanding the former rather than the latter.

However, while tornado charts show which of all uncertainties in a model are the greatest drivers of risk in the output, they do not help in understanding what values the output may give for certain values of the inputs, for example, what the profits/month would be if the consumer sensitivity is very low (e.g., at its 1-percentile) or very high (e.g., at its 99-percentile). In addition, there may be additional parameters that could affect model outcomes than solely our uncertain distributions. In the case of MoviePass, the movie admissions price also influences profitability, and we explored that in Chapter 2 by using a simulation table. Now, we will use Advanced Sensitivity Analysis to examine the effect of uncertain inputs and parameters, such as the movie admission price (which MoviePass will have some control over), to simultaneously analyze and compare their impacts on model outputs.

MoviePass6a.xlsx has our basic uncertain MoviePass model (using the initial $9 admission price). Advanced Sensitivity Analysis is found under Advanced Analyses on the @Risk toolbar. The dialog asks for a cell to monitor—we choose our total profitability (per month) cell, J28. Within the Advanced Sensitivity Analysis window, we then add our two distributions, increase in movie going (E12) and consumer sensitivity (E13), and our admissions price cell (D5). Each is added by selecting the Add button in the dialog. To add a cell rather than a distribution, click the dropdown arrow and change from "distribution" to "cell." There are a number of choices for values for each type—we will use the default percentiles for the distributions, and we enter the four admissions prices ($6, $7, $8, $9) as table values for the cell. By using the same four admissions price values that we used in the simulation table in Chapter 2, we will actually be conducting the same analysis here, but we will see how the output we obtain differs. Figure 6.1 shows the completed Advanced Sensitivity Analysis dialog.

There are several options for choosing outputs to obtain. After setting these, choose Analyze in Figure 6.1. @Risk runs 18 separate Monte Carlo simulations (each of these 18 times, it will run 10,000 iterations) and produces a workbook with a number of graphs, tables, and views of the output. The Tornado Graph is Figure 6.2

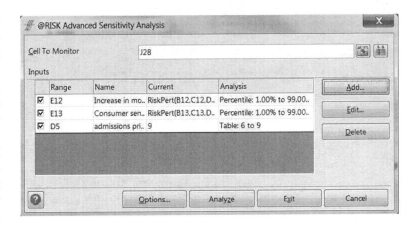

FIGURE 6.1
Advanced Sensitivity Analysis Setup

is just like the Tornado Graph from Chapter 2, except that all three of our inputs are shown. In other words, only two of the @Risk distributions are shown, and the admission price is shown, while in Chapter 2 all @Risk distributions were included in the tornado chart).

Figure 6.2 shows us that the range in possible admissions price is nearly as influential as the uncertainty in the consumer sensitivity on MoviePass profitability. A spider chart (called Percentage Change Graph) shows a different and more detailed view of the relative importance of the three inputs. In the additional tabs, the results of the Advanced Sensitivity Analysis explore the impact of each input separately, showing cumulative distributions and box plots (MoviePass6b.xlsx is a workbook containing all of this output).

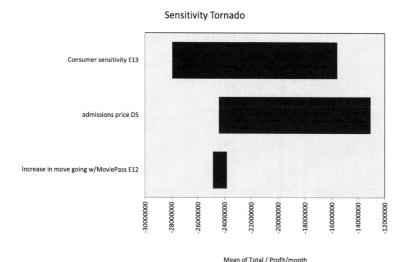

FIGURE 6.2
Tornado Graph for Advanced Sensitivity Analysis

Most of this analysis would also be available using simulation along with a simulation table, but especially when a model has many uncertain inputs, an Advanced Sensitivity Analysis can automate getting some of the outputs. The one additional feature available here is the ability to compare the influence of both uncertain variables described by distributions and other parameters represented in cells. This capability is useful when a key input is a decision variable (such as how to price MoviePass subscriptions—a question that could be explored with optimization, which we explore in Chapter 7). In the case of the movie admissions price, it is uncertain in that MoviePass was hoping to obtain reduced admissions prices from theaters, but instead of using a distribution to represent this, it is more insightful to either use a simulation table to explore its impact, or use advanced sensitivity analysis. The advantage of the latter approach is that we can see the relative importance of a decision variable compared with the uncertain components of our model.

6.2 Stress Testing

Tornado and Spider graphs permit us to examine the impacts of uncertain inputs on model outputs, but the analysis may not be as granular or as specific as we'd like. Tornado graphs split the input distributions into deciles, showing what happens to the output when an input is in each 10% of the varying input distributions. In other words, when a specific input is in its lowest decile, or in the second decile, what is the value of the output? A tornado chat then only shows the value of the output for the highest and lowest decile. Spider charts show the value of the output for all 10 deciles and can even be adjusted to show more than these 10 "buckets." For example, running the MoviePass simulation, choosing Spider Graph from the bottom buttons, and adjusting the settings to show 20 scenarios, results in Figure 6.3.

While tornado and spider graphs can provide this increased granularity, sometimes we are more interested in exploring how the outputs are affected when the inputs are at some "extreme value" or, in other words, if the inputs are at their tails of our input distributions. Stress testing allows us to do this. We can focus on a particular range of the input distribution—often this will be either the low or high end ("tail") of the distribution, although there may

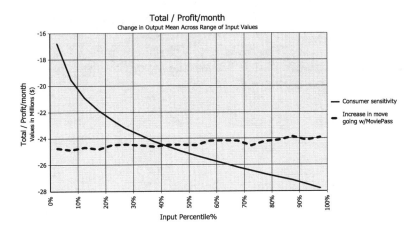

FIGURE 6.3
Spider Graph

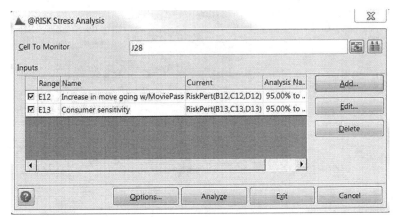

FIGURE 6.4
Stress Testing Setup

be instances where a different range could be of interest. This choice will depend on the context of the decision problem. For example, in the case of MoviePass, we might be particularly interested in what would happen if our consumers end up being at the extreme of how sensitive we thought they are, both in terms of subscription behavior and movie going increases, and how this affects profitability.

Stress Testing is found under Advanced Analyses on the @Risk toolbar. The dialog asks for a cell to focus on—we will use Total profitability (J28). We then add our two input distributions (only distributions can be used in Stress Testing). For each, a choice must be made of what part of the input distribution to explore, or "stress." We choose to stress the high end of both input distributions, representing a very high increase in movie going with MoviePass and a very high consumer price sensitivity. Our completed dialog is shown in Figure 6.4.

There are options that can be set, the most important of which is whether to stress each input in its own simulation (choose separate). Then, selecting "Analyze" produces a workbook with a number of views of the results (MoviePass6c.xlsx). Figure 6.5 shows box plots for the baseline and each input simulated in the top 5% of its range.

Figure 6.5 shows varying the increase in movie going to the upper range of its distribution has little impact, relative to the baseline, on profitability. Stressing this input does not matter much. On the other hand, stressing consumer sensitivity to its upper ranges profoundly decreases profitability. There is not much difference between the 95th and 100th percentiles, but the outcomes universally result in very low profitability. Much like the other sensitivity analyses we have done for this model, consumer sensitivity is a key input for MoviePass profitability. Stress testing allows us to focus on particular parts of this input distribution.

Additional Useful Techniques

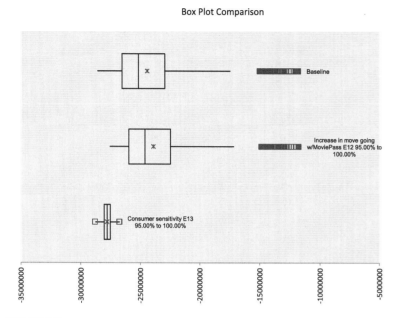

FIGURE 6.5
Stress Analysis Box Plots

6.3 Non-Parametric Distributions

There are times that no parametric distribution will provide a good description (i.e., no good fit) of the data. This is often the case for multimodal distributions. MPG6.xlsx contains data on the miles per gallon for the 3884 hybrid vehicles that were part of the National Transportation Survey in 2009.[1] The data has been sorted, and a histogram of the data, Figure 6.6, reveals the problem.

Suppose you want to model the actual fuel efficiency of hybrid vehicles, as actually driven by their owners. It is not likely that any parametric distribution will fit this data well. The multimodal distribution (three modes in this case) probably results from some hybrid vehicles being driven mostly in city traffic (where the electric engine is predominantly used), some driven mostly on highways (where the gasoline engine is primarily used), and some driven in a mixture of city and highway traffic. Figure 6.7 shows the "best-fitting" distribution for this data.

The Triangle distribution is the best fit according to the AIC, but it is not a good description of the data. We can do much better using either of two approaches. One is to fit a non-parametric distribution to the data, and the other is to use @Risk's Splice Distribution (twice in this case, to capture the three modes of the distribution). We will demonstrate both approaches, along with the best-fitting parametric distribution, and compare the results.

Figure 6.8 shows the model part of our spreadsheet. To use the non-parametric RiskCumul distribution, we first create the cumulative probability distributions in Columns F and G by using the Percentile function in Excel. Column E establishes the percentiles and Column F calculates those percentiles in the data. The cumulative probabilities are estimated in Column G using the approximation that the probability for each range equals the number of data points divided by the number of data points plus 1.[2]

The RiskCumul distribution asks for minimum and maximum values (F1 and F22, although you can use values outside the range of the data), the values to be simulated (F2:F21), and the cumulative probabilities attached to these values (G2:G21). Cell C2 contains the RiskCumul distribution that results.

The Splice distribution (RiskSplice) is found under Insert Function–Special on the @Risk toolbar. It requires two distributions that are connected via a splice point. In this case, there are two splice points (C3 and C4), which were selected visually from Figure 6.6. Then distributions were fit for each of the ranges of the data that fall below the first splice point, between the two splice points, and above the second splice point. To capture the entire multimodal distribution, we splice the first two distributions, and then splice the result to the third distribution. The final result is in cell C10. Figure 6.9 shows the simulation results for the three distributions.

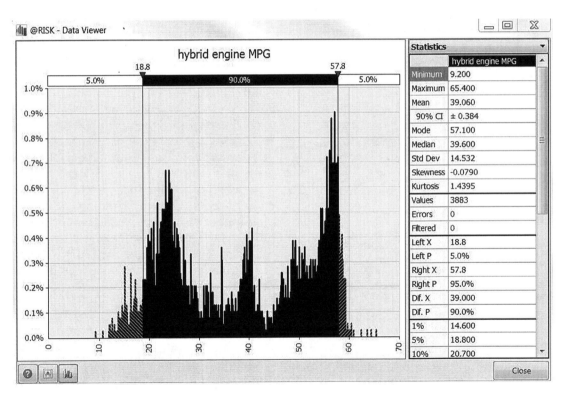

FIGURE 6.6
Miles per Gallon for Hybrid Vehicles

Additional Useful Techniques

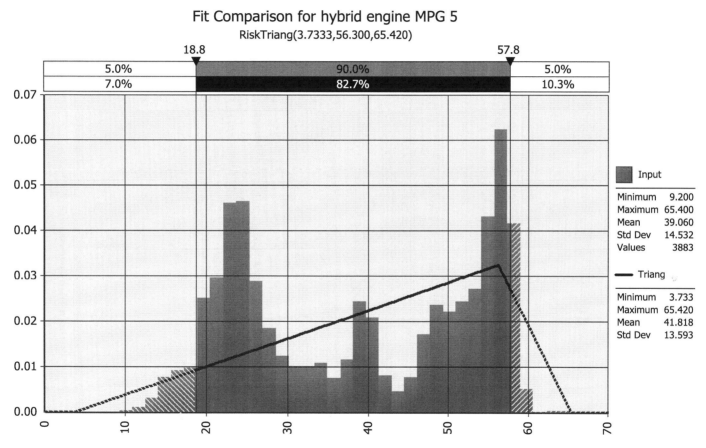

FIGURE 6.7
Distribution Fit for the MPG Data

	A	B	C	D	E	F	G
1	hybrid engine MPG	Best single fit	=RiskTriang(3.7333,56.3,65.42)		min	=MIN(A2:A3884)	cumulative prob
2	9.2	Non-parametric	=RiskCumul(F1,F22,F2:F21,G2:G21)		0.05	=PERCENTILE($A$2:$A$3884,E2)	=E2*(COUNT($A$2:$A$3884)/(COUNT($A$2:$A$3884)+1))
3	10.7	splice point 1	34.3		0.1	=PERCENTILE($A$2:$A$3884,E3)	=E3*(COUNT($A$2:$A$3884)/(COUNT($A$2:$A$3884)+1))
4	11.9	splice point 2	43		0.15	=PERCENTILE($A$2:$A$3884,E4)	=E4*(COUNT($A$2:$A$3884)/(COUNT($A$2:$A$3884)+1))
5	12	best fit <34.3	=RiskNormal(23.736,4.4623)		0.2	=PERCENTILE($A$2:$A$3884,E5)	=E5*(COUNT($A$2:$A$3884)/(COUNT($A$2:$A$3884)+1))
6	12	best fit >34.3 <43	=RiskTriang(33.3774,40,43.2813)		0.25	=PERCENTILE($A$2:$A$3884,E6)	=E6*(COUNT($A$2:$A$3884)/(COUNT($A$2:$A$3884)+1))
7	12.1	best fit >=43	=RiskExtvalueMin(55.4503,3.2122)		0.3	=PERCENTILE($A$2:$A$3884,E7)	=E7*(COUNT($A$2:$A$3884)/(COUNT($A$2:$A$3884)+1))
8	12.1				0.35	=PERCENTILE($A$2:$A$3884,E8)	=E8*(COUNT($A$2:$A$3884)/(COUNT($A$2:$A$3884)+1))
9	12.2	1st spliced distribution	=RiskSplice(C5,C6,C3)		0.4	=PERCENTILE($A$2:$A$3884,E9)	=E9*(COUNT($A$2:$A$3884)/(COUNT($A$2:$A$3884)+1))
10	12.3	Final spliced distribution	=RiskSplice(C9,C7,C4)		0.45	=PERCENTILE($A$2:$A$3884,E10)	=E10*(COUNT($A$2:$A$3884)/(COUNT($A$2:$A$3884)+1))
11	12.6				0.5	=PERCENTILE($A$2:$A$3884,E11)	=E11*(COUNT($A$2:$A$3884)/(COUNT($A$2:$A$3884)+1))
12	12.6				0.55	=PERCENTILE($A$2:$A$3884,E12)	=E12*(COUNT($A$2:$A$3884)/(COUNT($A$2:$A$3884)+1))
13	12.6				0.6	=PERCENTILE($A$2:$A$3884,E13)	=E13*(COUNT($A$2:$A$3884)/(COUNT($A$2:$A$3884)+1))
14	12.7				0.65	=PERCENTILE($A$2:$A$3884,E14)	=E14*(COUNT($A$2:$A$3884)/(COUNT($A$2:$A$3884)+1))
15	12.9				0.7	=PERCENTILE($A$2:$A$3884,E15)	=E15*(COUNT($A$2:$A$3884)/(COUNT($A$2:$A$3884)+1))
16	13				0.75	=PERCENTILE($A$2:$A$3884,E16)	=E16*(COUNT($A$2:$A$3884)/(COUNT($A$2:$A$3884)+1))
17	13				0.8	=PERCENTILE($A$2:$A$3884,E17)	=E17*(COUNT($A$2:$A$3884)/(COUNT($A$2:$A$3884)+1))
18	13.2				0.85	=PERCENTILE($A$2:$A$3884,E18)	=E18*(COUNT($A$2:$A$3884)/(COUNT($A$2:$A$3884)+1))
19	13.6				0.9	=PERCENTILE($A$2:$A$3884,E19)	=E19*(COUNT($A$2:$A$3884)/(COUNT($A$2:$A$3884)+1))
20	13.7				0.95	=PERCENTILE($A$2:$A$3884,E20)	=E20*(COUNT($A$2:$A$3884)/(COUNT($A$2:$A$3884)+1))
21	13.8				1	=PERCENTILE($A$2:$A$3884,E21)	=E21*(COUNT($A$2:$A$3884)/(COUNT($A$2:$A$3884)+1))
22	13.8				max	=MAX(A2:A3884)	
23	13.8						

FIGURE 6.8
Non-Parametric and Splice Distribution Models

Comparing Figure 6.9 with Figure 6.6 shows that both the non-parametric and Splice distributions are capable of recovering the multimodal nature of the data, while the parametric best-fitting distribution is not. However, the cumulative distribution provides a more accurate description of the data—and can be made more accurate (by using more granular percentiles). The drawback to using the cumulative distribution is that it will only simulate what is observed in the data. If the tails of the distribution are of most interest, then there may be few (or no) observations in those ranges.

Then, using parametric distributions can simulate values not present in the data—the splice distribution provides one way of doing this (provided that reasonable distributions can be found to fit various ranges in the data).

When fitting distributions to data, it is important to visually inspect the fit. In cases like this, where no single parametric distribution captures the shape of the data well, then these alternative methods are available to produce more accurate representations of the data.

Additional Useful Techniques

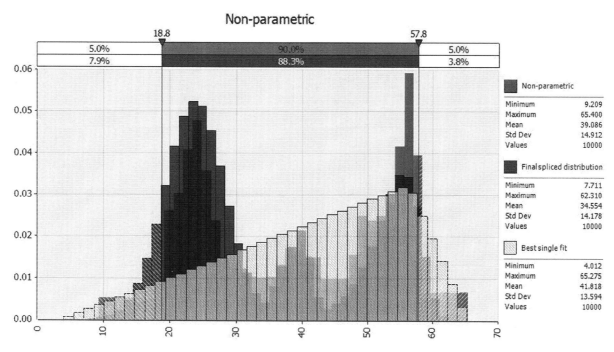

FIGURE 6.9
Comparison of Non-Parametric and Splice Distributions

6.4 Case: An Insurance Problem

A key reason that insurance companies exist is that people and companies are generally averse to risk. The essence of the business is that insurance is offered to a population of people that are potentially exposed to the risk of events that could have adverse financial consequences. By pooling the risks of a large number of people who are risk averse, the insurance company can offer an insurance policy at a price to the client that will cover its expected losses, administrative costs, and often even a profit, and clients will be willing to pay this premium to avoid the financial consequences of their individual potential losses. Of course, things get complicated when we consider that individuals vary in their riskiness and their attitudes to risk, and that the offer of voluntary insurance may itself affect the risks people take (this is known as moral hazard), as well as the likely purchasers of an insurance policy (i.e., adverse selection). These are all interesting complications, but at its essence, for an insurance company to be able to offer and price insurance policies and understand its risks, it needs to be able to estimate both the *frequency* of losses and the *severity* of these losses when they occur.

Our initial example is based on California auto insurance data for the 2010–2014 time period.[3] We will ignore details such as different risk categories and focus on one type of damage: collision damage. According to the California Department of Insurance, the probability[4] of a collision claim was 0.0604 per collision damage insurance policy per year, and the average damage claim was $3332 with a standard deviation of $2742.[5] We further assume that the administrative costs plus the profits) are 10% of the expected losses of each policy. Then, assuming that there is no further data on how risks vary across individuals (e.g., due to age, driving records, etc.), and that each individual's risk of a claim is independent, what annual collision premium must be charged for this insurer to breakeven?

6.5 Frequency and Severity

The circumstances in this problem are not unique to the insurance industry. The essential features are that there is an unknown frequency of something occurring, and every time it occurs, the impact has a certain (unknown) severity. There are numerous examples of situations where both frequency and severity distributions play an important role in understanding risk.

Film studios generally have a number of films in production, only some of which will be released to theatres (frequency). The released films will always vary in their box office receipts (severity). The same is true for publishers, music labels, pharmaceutical research and development, and drilling oil wells. Extended warranties offer consumers protection against the possible need for repair (at an unknown frequency as well as the varying size/costs of these repairs (severity). A gambling casino faces uncertainty in the number (frequency) of winners as well as the variability of the size (severity) of the wins. The exercises at the end of this chapter further illustrate the diverse set of situations that share the common features of an unknown frequency and severity.

Both frequency and severity may be modeled using different probability distributions. But care must be taken combining these together to estimate the total costs, or a mistaken risk assessment will result. We will illustrate this by returning to the insurance problem. The insurance problem is conceptually shown in Figure 6.10.

To estimate the losses the insurance company will need to pay out (either total losses or per policy holder), we need to estimate how many claims there will be and, when claims occur, how large each will be. For now (we'll relax this assumption in a bit), we assume that each insured person will experience either no collision claim or exactly one claim during a 1-year period, with a probability of 0.0604. This is modeled for each person as a Bernoulli distribution, shown in Figure 6.11.

Bernoulli distributions apply whenever there are two possible outcomes, a 0 or a 1, and a probability is attached to each of those outcomes (the probability for a 1 is commonly referred to as the "probability of success" or "p"). For our first model, we will assume

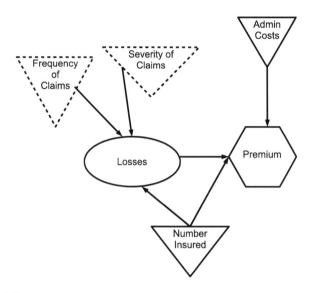

FIGURE 6.10
Insurance Visualization

that there are only 100 insured drivers (i.e., policies) and that the severity of claims follows a Lognormal distribution, as shown in Figure 6.12.

Lognormal distributions are one of several (severity distributions) that can be used to model insurance losses—note how the distribution is positively skewed. Most losses are relatively small, but occasionally a large loss occurs, which generates a long right tail of the (lognormal) severity distribution. Figure 6.13 shows our initial model (we have hidden the rows for drivers 4 through 98 and placed the final calculations at the top of Columns E and F for ease of display—the spreadsheet is available as Insurance6a.xlsx). This is a straightforward model where each driver is assigned a Bernoulli distribution to determine if there is a claim and a Lognormal distribution to determine the claim amount (severity) when a claim occurs. Note that we model the possible size of a claim whether a driver has a claim or not—but the actual total claims only count claim amounts for drivers where Column B shows that a claim has occurred (i.e., a 1).

The example model is using the Excel option to Show Formulas.[6] If you display the formula results rather than the formulas, and press the Dice (Random) button on the @Risk toolbar, you will see that most drivers do not file claims in any individual trial of a simulation. In fact, as you would expect, typically around 6 of the 100 drivers will file a claim (given it's a 0.0604 probability). Running 10,000 simulations (with a fixed seed value of 1) provides results for the total claim amount (including admin costs and profits) shown in Figure 6.14.[7]

Again, notice the "fat" right tail of potential total claims, indicating that there is potential for very high claim amounts, but that the probability is relatively low. The total expected claim amount plus administrative costs and profits from selling the 100 insurance policies is $222.62 per policy[8] (found on the right side of the @Risk results window—and which can be added to the histogram using the graph options dialog) and the 90% confidence interval shown in Figure 6.14 ranges from $64 to $434. Typically, an insurance firm would, however, also be very interested in understanding the upper tail of the total claim distribution to better understand how much capital they would have to reserve in order to be (for example) 99.55% or 99.9% confident that the amount reserved is sufficient to cover the losses in a particular year. Thus, we could also examine some of the upper percentiles (the 99.9th percentile is $719 per policy).

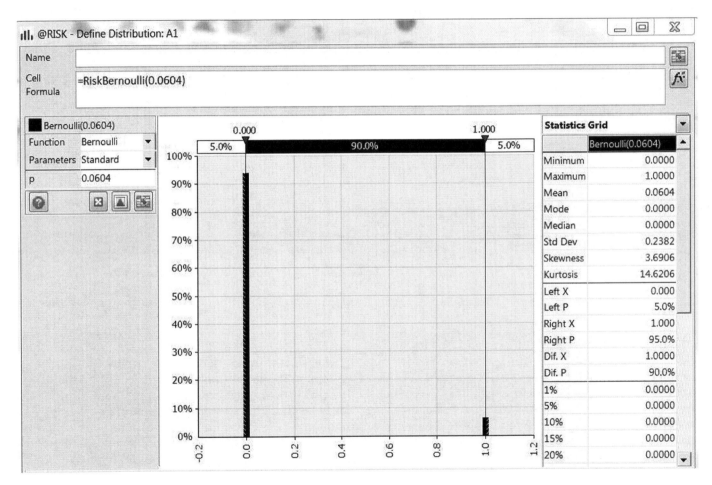

FIGURE 6.11
Bernoulli Distribution for Claim Frequency

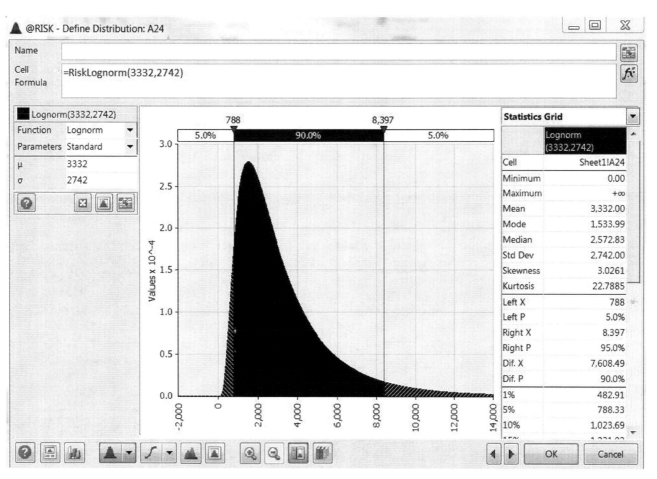

FIGURE 6.12
Lognormal Severity Distribution

	A	B	C	D	E	F	G	H	I	J	K
1	Parameters				Total Claims	=SUM(D10:D109)					
2	Number insured	100			Admin Costs	=B6*F1					
3	Claim frequency	0.0604	Bernoulli Distribution		Total Cost	=F1+F2					
4	Severity mean	3332	LogNormal Distribution		Premium	=RiskOutput()+F3/B2					
5	Severity std dev	2742			99.9th percentile	=RiskPercentile(F4,0.999)					
6	Administrative Costs	0.1									
7											
8	Calculations					Alternative Calculation					
9	Driver/Insured	Claim?	possilbe amount	actual claim		Number of Claims	Claim Amount	Total Claims	Admin Costs	Total Cost	Premium 2
10	1	=RiskBernoulli($B$3)	=RiskLognorm($B$4,$B$5)	=B10*C10		=RiskBinomial(B2,B3)	=RiskLognorm(B4,B5)	=F10*G10	=B6*H10	=H10+I10	=RiskOutput("Premium 2")+J10/B2
11	2	=RiskBernoulli($B$3)	=RiskLognorm($B$4,$B$5)	=B11*C11							
12	3	=RiskBernoulli($B$3)	=RiskLognorm($B$4,$B$5)	=B12*C12							
108	99	=RiskBernoulli($B$3)	=RiskLognorm($B$4,$B$5)	=B108*C108							
109	100	=RiskBernoulli($B$3)	=RiskLognorm($B$4,$B$5)	=B109*C109							

FIGURE 6.13
Initial Insurance Model

Of course, insurance companies typically have many more than just 100 policies, and therefore we would like to examine a larger model. We would also like to relax some of our assumptions, but it would be nice to find a simpler way to model this problem first. Just imagine trying to change the number of insured drivers to 500,000—it is not complicated to copy the rows further down (and adjust the summation formula), but it is unwieldy, unnecessary, and makes for a much slower model to simulate.

A little probability theory helps: if we have a number (n) of independent Bernoulli trials, the number of successes (out of n trials) will follow a Binomial distribution. The Binomial distribution is based on the number of trials and the (constant) probability of success per trial. For our 100-policy model (we will increase the number of policies shortly), the Binomial distribution is shown in Figure 6.15.

As Figure 6.15 shows, the highest probability is to get around 6 claims out of 100 policies (recall that the probability of a claim is .0604), but 5 and 7 are also fairly likely outcomes. Notice that we virtually always have at least 1 claim, and almost never will get more than 15 claims from 100 people. By using the Binomial distribution, we can simplify our model, as shown in Figure 6.13, cells F10:K10.

This is a much simpler spreadsheet, indeed! We only need one cell to simulate the number of claims out of 100 people, so we replace 100 rows of our spreadsheet with a single row. But, be careful, there is a problem—we mark AverageClaim2 as a second @Risk output and show the simulation results, overlaying our two simulation output cells as shown in Figure 6.16. Note that AverageClaim is the same distribution as in Figure 6.14, and AverageClaims2 is the total claims calculated using the Binomial distribution (instead of 100 Bernoulli distributions).

It is easy to see that the distributions look quite different—our simpler model (AverageClaim2) produces much more variable (wider) results, although the mean values appear almost identical. What has gone wrong?

The problem is that our second and smaller model is correctly modeling the frequency—the number of claims out of 100 people,

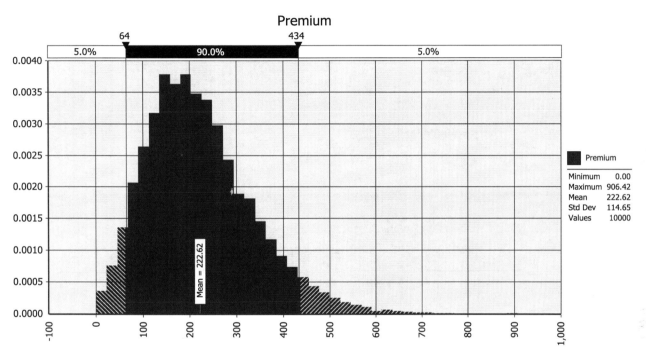

FIGURE 6.14
Average Claim Size and Expenses per Policy

but it is not correctly capturing the severity distribution, that is, the variability in the size of claims. By using a single cell for the Claim Amount (the cell with the Lognormal distribution, G10), our model is assuming that *every* claim that we simulate with the frequency distribution has the exact same claim size (i.e., the same severity). Imagine, for example, a scenario in which eight claims are simulated, and the Lognormal distribution (G10) simulates a severity of $5230. Now, in cell H10, the model multiplied 8 and $5230, but this implies that all of the eight claims had a claim size of $5230. In reality, if eight claims were filed, they would all have different claim sizes.

In other words, AverageClaim2 is simulating an unrealistic situation where every claim that may be filed for the 100 drivers

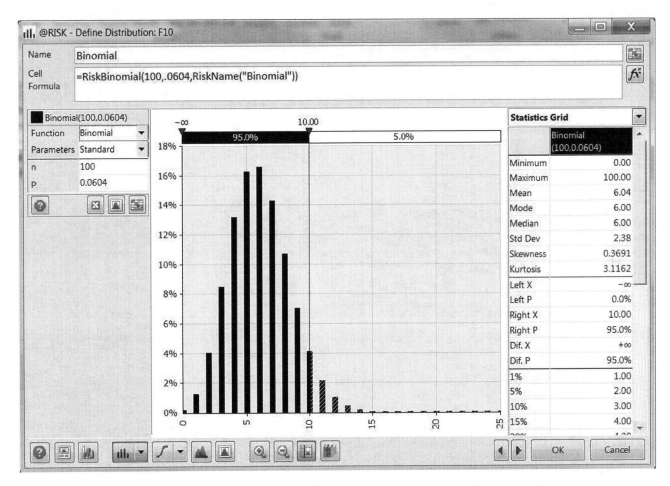

FIGURE 6.15
Binomial Distribution for the Number of Claims from 100 Policies

Additional Useful Techniques

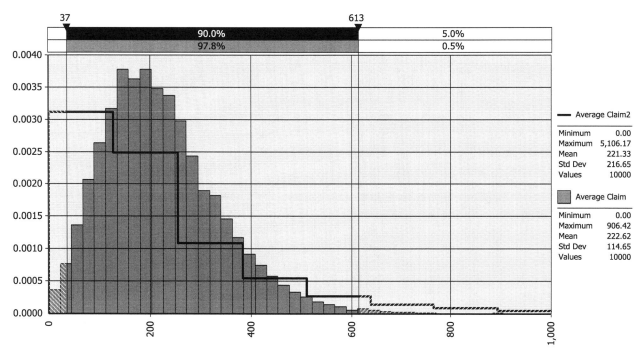

FIGURE 6.16
Comparison of the Long and Short Insurance Model Results

during the year has an identical severity. The probability of this happening is vanishingly small. The original model correctly permitted each claim to vary independently of the others. This results in much less overall uncertainty (i.e., a narrower distribution) in the resulting total claims costs, since some large claim amounts will be offset by others with small claim amounts. The simple (smaller) model incorrectly does not permit large and small claims to offset each other—all claims are forced to be the same size in each simulation trial.

Fortunately, there is a way to fix this problem, and at the same time keep the model simple, which is through the use of *compound distributions*.

6.6 The Compound Distribution

The compound distribution (in @Risk called the RiskCompound) is found under Insert Function–Special on the @Risk toolbar. This distribution is designed specifically for frequency-severity problems, where there is a discrete frequency distribution that needs to be combined with a continuous severity distribution such that each discrete event has its own random severity level. The RiskCompound function sums (or aggregates) a number of independent and identically distributed ("iid") random variables. For example, if drivers' collision damage distributions are iid, then the *total* damage of a portfolio of collision damage insurance policies is a random variable that is obtained by summing these individual distributions, one for every accident. We could model each distribution separately and then add them together as we did in the original model, or, alternatively, we can use the RiskCompound function to do this for us.

The RiskCompound function requires the input of a frequency and severity distribution. In addition, the RiskCompound function has optional inputs that permit the inclusion of a deductible and/or maximum liability amount (these are common features for insurance policies, which we will not include in our current model). We adapt our insurance model to include the same Binomial and Lognormal distributions that we used to estimate AverageClaims 2, but using the RiskCompound distribution to estimate a new Premium 3 (we now refer to these as premiums since they include average claims as well as administrative costs), as shown in Figure 6.17 and spreadsheet Insurance6.b.xlsx.

Running 10,000 simulations, the results are shown in Figure 6.18. Notice that the distributions for Premium and Premium3 are very similar, but Premium2 has a much wider (and incorrect) distribution. Unlike our previous incorrect attempt to reduce the model to a single row (Premium2), this single row (Premium3) correctly models the insurance problem.

To fully appreciate the power of the RiskCompound function, we will now examine the total claim size distribution for an insurance company with many policies (1,000,000), and with some additional realism. In the previous analysis, we assumed that each policy could either have zero or one claims. However, in reality, some policies may experience more than one claim, and the Bernoulli distribution only permits outcomes of zero or one claim for each driver. Therefore, we will change the Claim Frequency distribution from a Binomial distribution to a Poisson distribution (with the rate[9] equal to the [number of policies] × [claim rate per policy per year]). In many models structured with a frequency and severity, the frequency distribution is modeled as a Poisson distribution rather than a Binomial.[10] In our example, the number of events can be thought of as either a number of trials (100 policies) with a specific probability of a claim occurring or not—in which case we would use Binomial (n = 100, p = 0.064)—or as a number of events with an average rate of occurrence of 0.0604 per year—in which case we would use Poisson (100*0.0604). Because in this example, the rate per year is small, it will be rare to have two claims per year for a policy, and therefore the Binomial and Poisson distributions will be very similar and give similar results in this case. If it is possible to have more than one claim per policy per year, then using the Poisson distribution is correct (and the Binomial approach is incorrect, but a close approximation because 0.0604 is a low probability).

We also include a deductible of $500 per claim and a maximum liability of $50,000 for comparison with policies without any deductible or maximum liability limit.

Figure 6.19 and spreadsheet Insurance6c.xlsx show our new model. The results of this model are shown in Figure 6.20. This example is also a great demonstration of how spreading risk across a large population of policies can drastically reduce the risk assumed by

	E	F	G	H	I	J	K
1	Total Claims	=SUM(D10:D109)					
2	Admin Costs	=B6*F1					
3	Total Cost	=F1+F2					
4	Premium	=RiskOutput()+F3/B2					
5	99.9th percentile	=RiskPercentile(F4,0.999)					
6							
7							
8		Alternative Calculation					
9		Number of Claims	Claim Amount	Total Claims	Admin Costs	Total Cost	Premium 2
10		=RiskBinomial(B2,B3)	=RiskLognorm(B4,B5)	=F10*G10	=B6*H10	=H10+I10	=RiskOutput("Premium 2")+J10/B2
11							
12		Risk Compound Approach	Admin Costs	Premium 3			
13		=RiskCompound(F10,G10)	=B6*F13	=RiskOutput("Premium 3")+(F			

FIGURE 6.17
Use of the RiskCompound Function

the insurance company. Compare the 90% confidence interval of the AverageClaim distribution in Figure 6.14 (ranging from $64–$434) to the 90% confidence interval (without the deductible and limit) with 1,000,000 policies (ranging from $219 to $223).[11] The reason the latter interval is so narrow is that with 1,000,000 policies (assuming that claims are independent), there are 10,000 times as many policies and therefore a few more or fewer incidents does not have much effect on the average claim size anymore. In fact, more policies tend to cancel each other out, producing a mean loss per insured that spans a much smaller range of values. A much larger risk pool has made the average (and total) loss quite predictable.

We can also see the beneficial effects (from the insurance company's view) of deductibles and liability limits. The average claim is reduced by almost $33 (so drivers will likely get some of these benefits).

6.7 Uncertainty and Variability

A distinction is often made between uncertainty and variability. A common distinction is that uncertainty has to do with our lack of knowledge (or ignorance) of parameters' values, and can typically be reduced through acquiring more data, while variability is an attribute of the system that is not reduced with more information. Both are modeled with probability distributions, but it is important to understand the distinction between both uses of distributions. For example, the number of insurance claims will vary from year to year, and we can represent this year-to-year variability with probability distributions. However, because the actual average rate at which claims occur per policy per year is typically estimated from a limited dataset, this rate has uncertainty, which can also be represented with

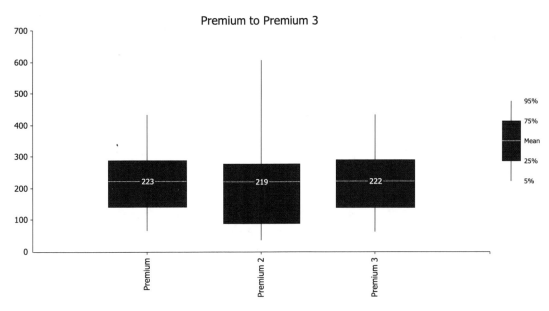

FIGURE 6.18
Insurance Results with the Compound Distribution

a probability distribution. Thus far, our models have assumed that the rate of claims is known with certainty (exact 0.0604 claims per policy per year). Suppose, however, that this is not the case—but that, in fact, the rate of claims is not known exactly. Suppose this parameter uncertainty can be modeled with a Normal distribution, with mean 0.0604 and a standard deviation greater than zero.[12] Uncertainty about this true parameter value will result in more risk for the insurer. Indeed, it does, as we now illustrate with an example.

Figure 6.21 shows the model (Insurance6d.xlsx; 1,000,000 policy holders, $500 deductible, and $50,000 maximum limit). The model results are shown in Figure 6.22. The mean premium is very similar for the two cases. However, without uncertainty about the rate of claims, the confidence interval around the average claim per policy is very narrow, with a small standard deviation ($1.07) for 1,000,000 drivers. Including uncertainty in the rate dramatically expands the range of this confidence interval. The insurer's risk exposure is much larger: the standard deviation triples. Our use of the Normal distribution (with standard deviation of 0.001) has substantially increased the uncertainty into the actual frequency of claims, and therefore the average claim per policy.

	A	B	C
1	Parameters		
2	Number insured	1000000	
3	Claim frequency	0.0604	Bernoulli Distribution
4	Severity mean	3332	LogNormal Distribution
5	Severity std dev	2742	
6	Administrative Costs	0.1	
7	Frequency Distribution	=RiskPoisson(B2*B3)	
8	Severity Distribution	=RiskLognorm(B4,B5)	
9	Risk Compound Approach	Admin Costs	Premium 3
10	=RiskCompound(B7,B8)	=B6*A10	=RiskOutput("Premium 3")+(A10+B10)/B2
11	With Deductibles and Max		
12	Deductible	500	
13	Maximum Liability	50000	
14	=RiskCompound(B7,B8,B12,B13)	=B6*A14	=RiskOutput("Premium 3 with Deductible & Limit")+(A14+B14)/B2

FIGURE 6.19
More Realistic Insurance Model

6.8 Bayesian Analysis

Bayesian approaches and methods to analyze data in order to support decision making are increasingly being used. The basic idea of Bayesian approaches is that we have current knowledge (or lack thereof, also called "uncertainty"), and when new/additional data is collected, we can use this data to update our knowledge A simple example will illustrate the basic thinking behind this approach and how it can be implemented in a simulation model.

Consider the case of the Zika virus. Exposure to Zika during pregnancy has been linked to many serious birth defects. One early study of 125 exposed women found 58 with adverse pregnancy outcomes.[13] A more recent study of 76 births found 42 with moderate or severe related complications.[14] While there have been many studies of Zika, let's assume these are the only two studies and we wish to estimate the prevalence of Zika-related birth defects in a population of 100,000 based on this new study.

A Bayesian analysis is based on the principle that we learn. The method utilizes Bayes rule, which shows how our knowledge (and lack of knowledge) is updated as information is gained. In summary, Bayes rule says that

New Knowledge = Prior Knowledge * Likelihood

This means that we take the prior knowledge and add to it (mathematically we multiply it by) the likelihood that we would get what we observed. The result of this multiplication is the updated

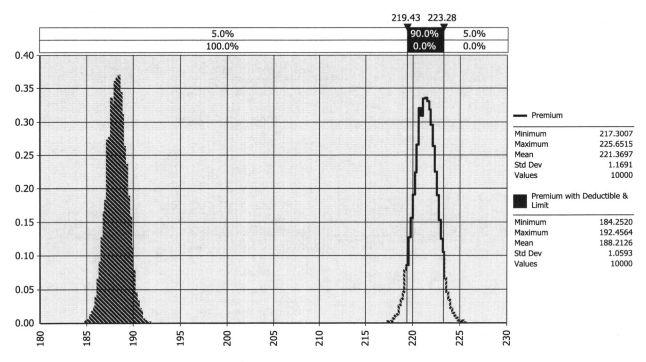

FIGURE 6.20
Insurance Results for 1 Million Drivers

knowledge (called the "posterior" probability). An important assumption of our analysis is that the underlying true rate of Zika-related birth defects is the same in each study—the evidence, however, differs.[15]

We can implement this for a simple example in a spreadsheet (Bayes6.xlsx). Figure 6.23 shows a section of the spreadsheet model where we construct the Bayesian estimates of the proportion. Since we are trying to estimate a proportion, you will see that the model is only evaluating all values between 0 and 1.

Row 2 of the spreadsheet contains the data from the prior study. For the new study results (42 birth defects out of 76 births), cells A6:D10 show the frequentist approach to estimate the standard error, which

Additional Useful Techniques

	A	B	C	D
1	Parameters			New Formulation
2	Number insured	1000000		
3	Claim frequency	0.0604	with uncertainty	=RiskNormal(B3,0.001)
4	Severity mean	3332		
5	Severity std dev	2742		
6	Administrative Costs	0.1		
7	Frequency Distribution	=RiskPoisson(B2*B3)	Poisson with uncertain frequency	=RiskPoisson(B2*D3)
8	Severity Distribution	=RiskLognorm(B4,B5)		
9				
10	With Deductibles and Max			
11	Deductible	500		
12	Maximum Liability	50000		
13	Compound Function	Admin Costs	Premium	
14	=RiskCompound(B7,B8,B11,B12)	=B6*A14	=RiskOutput("Premium with Deductible & Limit")+(A14+B14)/B2	
15	with uncertain frequency			
16	=RiskCompound(D7,B8,B11,B12)	=B6*A16	=RiskOutput("premium with uncertain frequency")+(A16+B16)/B2	

FIGURE 6.21
Insurance Model with and without Uncertain Frequency

treats the new study as the only information available. The standard deviation for the expected proportion of birth defects is the formula for the standard error of a proportion and results in a confidence interval (44%, 66%).

Instead, the Bayesian approach starts with prior knowledge for the true proportion with birth defects. The A21:E121 section of the spreadsheet (the first five rows of which are shown in Figure 6.23) is based on a *flat* prior—all possible percentage support levels (from 0 to 1) are equally possible. The spreadsheet also contains a section where the prior polling information is used to show the initial birth defect rate from the prior study (G21:K121). In both cases, the Likelihood function (Columns C and I) calculates a Binomial distribution showing the probability that the new birth defects (42 of the 76 births) could be at the level observed, as a function of each potential true proportion of birth defects. Then, in the Posterior columns, (Columns D and J) the prior knowledge and the likelihoods are multiplied. Finally, the posterior probabilities are normalized so that they add up to 1, which helps with graphically displaying several posterior distributions in one chart.

The section between Rows 13 and 17 of the spreadsheet collects the simulation results. The uncertainty around the proportion of birth

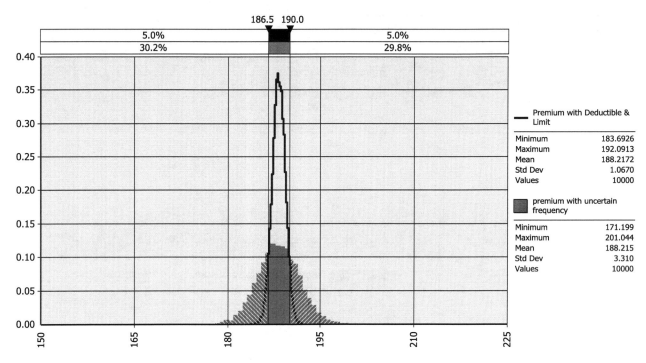

FIGURE 6.22
The Effect of Uncertainty in the Rate of Claims on Insurance Premiums

defects is simulated by a RiskGeneral distribution (which uses the Posteriors for the two situations of only considering the new study, or for considering both studies).[16] This estimated number of people in the total population of the jurisdiction is then multiplied by the low and high estimate of proportion of the population that will experience birth defects, which results in a Bayesian credibility interval (a sort of confidence interval, in this case 95%, for the Bayesian estimated birth defect rate).

Figure 6.24 shows both posterior credibility intervals for the proportion of birth defects.

The frequentist interval is only based on the latest study results and is centered on 55%. Notice that the credibility interval with

Additional Useful Techniques

FIGURE 6.23
Bayesian Model Setup

	A	B	C	D	E	F	G	H	I
8			Low	High				Low	High
9	Using confidence interval width formula	11.18%	0.441	0.664				44084	66442
10	Using inverse cummulative formula		0.441	0.664					
11									
12									
13	Bayesian analysis - with construction of the Bayesian posterior: confidence interval		Low	High				Low	High
14		Simulation							
15	With uninformed prior	0.509	0.440	0.660	<< need to run simulation to o			44035	65963
16									
17	Combined results	0.552	0.429	0.567	<< need to run simulation to o			42851	56652

FIGURE 6.24
Bayesian and Frequentist Confidence/Credibility Intervals

the uninformative prior is essentially identical to the frequentist confidence interval. However, the informative prior results in a narrower interval with the mean between the observed proportions in the two studies. The Bayesian credibility interval uses the prior study information—in this case that only 46% of births had defects. Figure 6.25 shows the prior and posterior distributions.

Thus, the new information considerably shifts the support toward the earlier defect rate of 46% (but not all the way, with the extent of the shift based on the sample sizes in the prior and current studies). If the prior study had also found a 55% birth defect rate, then both distributions would be centered on 55%, but the Bayesian interval would be narrower, since there would be a larger sample size supporting that estimate.

Next, we encourage you to put in some alternative study results into the input section of the model (cells B2:D3). For example, change the results or the size of the prior results, and see how that affects the estimate of the proportion. You will see that the smaller the prior

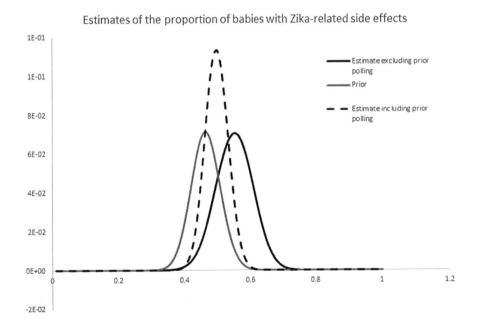

FIGURE 6.25
Prior and Posterior Distributions

study was in size, the greater the uncertainty (i.e., the uncertainty distribution will be wider) and the less weight/influence it will have on the combined results.

Another thing you'll find is that if the estimates in both studies are very different, the combined results will be somewhere "in between." One important aspect here is that a key assumption when performing the Bayesian analysis shown above is that when we're combining prior information with new data, we're still estimating the same parameters. In other words, if we can assume that the birth defect rate hasn't changed between the two studies, then combining them into one estimate is valid. However, if there was sufficient time or differing demographics between the studies, and the defect rate may differ, then using the prior study in this way would not be valid.

Bayesian analyses can get quite complicated, as, for example, the likelihood function may not be easy to compute. In our example, it was easily computable since it only required use of the Binomial distribution. Bayesian techniques are increasingly used in diverse fields such as finance, health, pharmaceutical research, and marketing. This chapter provided a brief illustrative example of how Bayesian techniques can be (and are) often included in Monte Carlo simulation models.

Exercises

Exercise 6.1 Hospital Bed Utilization

The Centers for Disease Control and Prevention reports that the average rate of discharge from hospital stays was 1139.6 per 10,000 people in 2010. The average length of stay was 4.5 days. Consider a medium-sized city with a population of 500,000 and five hospitals, each with 155 beds.

Beds can be utilized 100% without any downtime between patients. Assume the rate of daily admissions follows a Poisson distribution and the rate of admissions is evenly spread out over the year. The length of stay is a continuous variable and follows a Lognormal distribution (with the mean given above and standard deviation of 3 days).

a. What is the probability that this city has adequate hospital beds for the yearly demand for beds? What is the average annual occupancy rate as a percentage of the total beds in the city?

b. Assume that the number of available beds is affected by downtime between patients (and periodic maintenance). Over the course of the year, the number of available beds as a percentage of total beds follows a Pert distribution (minimum = 88%, mode = 91%, maximum = 93%). Recalculate your answer to part (a).

c. How frequently will occupancy rates greater than 80% be observed? Greater than 90%?

Exercise 6.2 Utility Company Repair Vehicles

Consider a utility company with a fleet of 50 repair vehicles. Each day, assume that the number of hours a vehicle will be used per trip follows a Lognormal distribution (with an average of 4 hours and a standard deviation of 2 hours). The number of repair trips taken in total for the company per day follows a Poisson distribution (with an average of 50 per day). Repair trips require travel, which follows a Lognormal distribution (with a mean of 20 miles and a standard deviation of 10 miles). The current truck fleet gets 15 miles per gallon, and the average fuel price is $3.50 per gallon.

a. Estimate the expected annual fuel costs for the repair fleet. Provide a 90% confidence interval for the annual fuel costs.

b. Comment on the current capacity and utilization of the repair fleet.

Exercise 6.3 Automobile Insurance: Moral Hazard and Adverse Selection

Data from an automobile insurance company shows that claims frequency and size are affected by what type of insurance is chosen by an individual.[17] The data is from an Israeli insurer for the years 1994–1999. Table 6.1 provides relevant data.

Model the claim frequency using a Poisson distribution (to allow for more than one claim per policy holder) and use a Lognormal distribution for the size of claims (with the mean given in Table 6.1 and standard deviation equal to the mean). Assume 1000 drivers are insured by the policies, in the proportions indicated in Table 6.1. Also assume there is no correlation between policy types. Build a simulation model to

TABLE 6.1

Insurance Policy Types

Policy Type	Average Premium	Deductible	Claim Frequency	Average Damage/ Claim	% Choosing
Regular	2800	1400	21.33%	11,433	76.72
Low Deductible	3640	840	27.76%	10,233	21.96
High Deductible	1960	2520	15.05%	13,600	0.74
Very High Deductible	1920	3640	11.37%	10,750	0.6

a. Estimate the contribution margin (revenue minus direct costs from claims, i.e., what is left to cover overhead costs and profits) per insured driver, for each policy type, and for the total of all drivers insured by this company.

b. See whether the differences in premiums appear to match the difference in expected costs across policies.

c. Determine the overall probability that this company will have a positive contribution. What level of contribution do you estimate there is a 95% likelihood of achieving or exceeding?

Exercise 6.4 Customer Loyalty Programs

Customer loyalty programs are designed to increase revenues from regular customers (and increase their retention).[18] Suppose grocery store customer purchasing behavior follows the pattern in Table 6.2.

Current company policy is to upgrade a customer to "Gold" status if their yearly purchases exceed $5000. They are demoted from Gold status when purchases fall below this amount. Gold status costs the company $400 in yearly benefits. Model the variability in the frequency of purchases using Poisson distributions. Model variability

TABLE 6.2

Loyalty Program Data

Customer Status	Average Number of Annual Transactions	Average Transaction Amount
Normal	39.5	$107.42
Upgraded to Gold	43.4	$122.57
Demoted from Gold	32.7	$92.21

in the level (severity) of purchases using Lognormal distributions (with means given in the table and standard deviations equal to 20% of these means).

a. For a typical year, estimate the probability of a customer in each status, staying in that status, upgrading to Gold, or demoting from Gold. What are your assumptions within this analysis?

b. Estimate the average gross profit (annual sales minus loyalty program cost) for each status, and comment on the profitability of this loyalty scheme for this grocer.

c. Would changing the threshold for Gold status to $4000 improve this loyalty scheme? Assume that customers have the same behavior (Table 6.2) after the change in threshold.

d. Perform a stress test to examine the most unfavorable circumstances for the original loyalty program (use the worst 5% of the distributions).

Exercise 6.5 Robocall Marketing

Robocalls have become a major problem for consumers and telecommunications companies. They represent the largest category of complaints filed with the Federal Communications Commission (FCC) and are the subject of several potential legislative acts. Estimates for January 2019 are that there were 4173 robocalls per second, or 250,391 per minute, or 21 spam calls for every person, in the United States alone.[19] Robocalls cost spammers very little, but not nothing. Estimates are that costs might be around 0.7 cents per minute. Estimates of consumer losses range from $350 million to $9.5 billion annually due to these calls.[20] Another estimate puts the average loss for victims at $357.[21] Another report claims that at least 30 customers in New York City lost $3 million due to one particular robocaller.[22] In 2016, one estimate put losses at $9.5 billion for 22 million Americans.[23] In addition, enforcement actions against robocallers have been difficult—the FCC has only collected $6790 in actual penalties despite doling out $208 million in fines.[24]

This is a rapidly evolving environment and there is much uncertainty about the actual rate of spam robocalling, their cost, and the response rates of, and damages to, victims.

a. Using the above data points and any other estimates you may need and can find, build a simulation model to estimate the aggregate profitability of the robocall scam industry. Explain all of your assumptions and your reasoning for selecting the distributions and parameters that you choose.

b. Among the potential "solutions" to the robocall problem are the following: technology to reduce the rate of robocalls getting through and increased probability of scam detection and punishment for scammers (you can use the fraction of fines collected as an estimate of detection and punishment if no other data can be found). Provide a preliminary model for these two types of solutions and use that model to describe the potential impact on robocall profitability. How much would the response rate need to be reduced in order to make robocalling unprofitable?

Exercise 6.6 Oil Spills

As the world depends on oil that is harder to reach (e.g., from Alaska, deep in the Gulf of Mexico, etc.), the threat of oil spills is ever present. The data in Table 6.3 is adapted from a report issued by the Bureau of Ocean Energy Management about oil spills in the US Outer Continental Shelf from 2001 to 2015.[25]

For the oil spill sizes, use Lognormal distributions (using the given means and medians, approximate these Lognormal distributions as closely as possible)—for the open-ended category, use 10,000 as the mean and 100,000 as the standard deviation). Model the number of spills of the various sizes using a Poisson distribution.

a. Build a simulation model for the total amount of oil spills for platforms and for pipelines (based on 8Bbbl of production). Provide 90% confidence intervals for each. What are the assumptions in this analysis?

b. What is the probability that the total pipeline spill volume will exceed the total platform spill volume?

c. Repeat part (b), but without the single large platform spill (the Deepwater Horizon spill).

Exercise 6.7 Supermarket Checkout

Suppose that customers arrive at a supermarket cash register at the rate of 45 per hour. Each customer takes an average of 2.85 minutes to check out, although this varies (it follows a Lognormal distribution with a mean of 2.85 minutes and a standard deviation of 3 minutes).

TABLE 6.3

Oil Spill Data

Spill Size (bbl)	# Spills		Bbl Spilled		Spill Rate (Based on 8Bbbl Production)		Average Spill Size (Barrels)		Median Spill Size (Barrels)	
	Platforms	Pipelines	Platforms	Pipelines	Platforms	Pipelines	Platforms	Pipelines	Platforms	Pipelines
≥1 to <5	333	59	670.5	137.2	49.4	9.6	2.0	2.33	2.5	4.3
≥5 to <10	62	18	431.0	121.1			7.0	6.73		
≥10 to <20	48	22	641.0	284.8	12.3	4.4	13.4	12.9		
≥20 to <50	50	13	1624.1	365.4			32.5	28.1		
≥50 to <100	32	8	2082	565.3	10.3	2.3	65.0	70.7	127	108
≥100 to <500	50	10	10,372	1952			207.4	195.2		
≥500 to <1000	10	3	6266	2493	1.3	0.4	626.6	831.0		
≥1000 to <2000	2	3	3066	4536	0.4	0.4	1533	1512	1572	1500
≥2000 to <3000	1	None	2000	None			2000	None		
≥3000 to <10000	None	None	None	None			None	None		
≥10000	1	None	4.9 million	None	0.1	None	4,900,000	0	4,900,000	None

a. Use distributions for frequency of checkouts and severity (duration of the checkout), and aggregate distributions to generate a histogram of the number of cashier minutes per hour that are required for customer checkout. How many cashiers would be necessary to have an 85% probability of meeting the demand during any hour? Assume that cashiers are available for all 60 minutes per hour and that there is no time required between customers.

b. Assume that the supermarket employs the number of cashiers you calculated in part (a). Create a histogram of the percentage of their time that is not occupied with customers.

Exercise 6.8 Snowfall at a Ski Area

Data for the years 2009–2017 for the Jackson Hole, Wyoming, ski area reveals that the average seasonal (October 1–April 15) snowfall is 1048 cm (413 inches), but ranged from a low of 735 cm to a high of 1562 cm—the data is in Exercise6-8.xlsx. Snowfall is critical to the ski industry. Your job is to provide a 90% confidence interval for the total yearly snowfall expected at Jackson Hole. Further investigation reveals that it snows on an average of 77 of the days during the 194-day snow season (use a Poisson distribution to model the number of days it will snow).

a. Produce a 90% confidence interval for average seasonal snowfall. What assumptions did you make?

b. What is the probability of getting a season with more snowfall than the highest recorded during the 2009–2017 decade (1562 cm)?

c. The number of days with snowfall was actually observed during this time period, but it is only an estimate of the true frequency of snow. Assume that this frequency is only an estimate and use a Poisson distribution for the number of days with snowfall each season. But assume the true frequency of snow (percent of days with snow) follows a lognormal distribution (with mean percent of days with snowfall as given from the data and standard deviation 6%). Answer parts (a) and (b) after incorporation of this uncertainty into your model.

Notes

1. This national survey is administered every 8 years. The 2017 survey did not report fuel efficiency, so we are using data from the 2009 survey.
2. If the dataset is small, then this additional setup of percentiles is unnecessary. Instead, each original data point is given the probability $=1/(1+n)$, where n is the number of data points, and these probabilities are then accumulated. For large datasets, this is more easily accomplished by breaking the data into percentiles, as shown in Figure 6.8.
3. The 2008 *California Private Passenger Auto Frequency and Severity Bands Manual*, Third Edition, Rate Regulation Branch, California Department of Insurance, April 20, 2018. The claim frequency is stated on the basis of exposure years, so this is not the same as the frequency of a claim per driver since some drivers will have more than one claim. We will examine this later in the chapter, but for now we assume that each driver has 0 or 1 claim per year.
4. This actually is a rate (and not a probability) per collision damage insurance policy per year, but we'll discuss this in Section 6.6 in the chapter.

5. The Third Edition of the manual did not report the standard deviation of the severity distribution, so we are using the standard deviation reported in the Second Edition. The mean values in both manuals were quite similar, so this may be a reasonable assumption.
6. This can be accomplished by using Ctrl+~ or by using the command in the Formula Auditing box under the Formulas tab in Excel 2007 or later.
7. It is more common to look at the total claim distribution, and this is what an insurance company would do. In this chapter, we express things on a per-policy basis because total claims depend on the scale of the model (the number of insured drivers), which we have arbitrarily assumed, and because it is an easier quantity for individual drivers to relate to.
8. You may have noticed that this is indeed very close to $0.0604*\$3321 + 10\% = \$221.38$. If we had run more trials, our Monte Carlo approximation would have gotten closer and closer to this amount.
9. The rate is also sometimes called lambda.
10. The Binomial and Poisson distributions are closely related, and sometimes the Binomial distribution can be approximated by a Poisson distribution. The Poisson measures the number of events within a range of exposure (e.g., over a certain time period), while the Binomial measures the number of successes out of a specified number of trials.
11. To put this in some context, the Insurance Information Institute reports the average collision policy in California was priced at $423.75 (www.iii.org). Perhaps administrative costs are higher than we have assumed—or the discrepancy between expected costs and policy rates is an indication of the adverse selection problem facing insurers. Collision insurance is optional, and we should expect drivers with higher-than-average expected claims to purchase policies.
12. A Normal distribution is assumed here, since normality is often assumed in statistics when estimating the uncertainty about a parameter. We will assume a standard deviation of 0.001 for this example (which is also sometimes called the standard error).
13. Brasil, P. et al. 2016. "Zika Virus Infection in Pregnant Women in Rio de Janeiro." *New England Journal of Medicine*, 375: 2121–2334.
14. Pomar, L. et al. 2018. "Maternal-Fetal Transmission and Adverse Perinatal Outcomes in Pregnant Women Infected with Zika Virus: Prospective Cohort Study in French Guiana." *BMJ*, 363: k4431.
15. Alternatively, consider the case of political polling. The Bayesian approach might use earlier polls, or polls from different locations, as prior information, with subsequent polls as new information used to update the earlier estimates. The implicit assumption is that the voter preferences have not changed from one poll to the other. Similarly, the two Zika studies referenced here were conducted in two locations at two points in time: our assumption is that the underlying rate of birth defects has not changed. More sophisticated analyses would be required if we believe the underlying probabilities have changed in order to model those changes.
16. The general distribution is a non-parametric distribution requiring only a set of possible values in ascending order and a set of posterior weights for each of those values. The general distribution will normalize the posterior probabilities. The model also uses the beta distribution to model the probabilities—often used to model uncertainty about a binomial probability, as it is a flexible distribution with a bounded range of values (in this case, zero to one).
17. The data is adapted from Cohen, A. 2005. "Asymmetric Information and Learning: Evidence from the Automobile Insurance Market." *Review of Economics and Statistics*, 87:197–207. We have simplified the data and made some additional assumptions to produce Table 6.1. The numbers are measured in New Israeli shekels.
18. For an investigation based on some real data, see Wagner, T., Hennig-Thurau, T. and Rudolph, T. 2009. "Does Customer Demotion Jeopardize Loyalty." *Journal of Marketing*, 73:69–85. That study suggests that customers do not increase their purchases when they receive "gold" status but significantly decrease purchasing behavior when being demoted from gold status. The latter effect is not surprising, but the former is. We have used hypothetical data for

what we believe is the more typical pattern of behavior, where gold status has some positive impact, but demotion does indeed cause purchases to drop.

19. From www.robokiller.com.
20. "Congratulations! You Lost." *Slate*, May 24, 2016, and "Don't Pick Up: How to Stop These Annoying Robocalls for Good," *USA Today*, April 20, 2018.
21. "Robocalls Spike in Tax Season. Government Shutdown Gave Scammers a Boost. So Guard That Info." *Clarion Ledger*, Feb. 8, 2019.
22. "By the Time You Finish This Article, 400K Americans Were Probably Robocalled." *NBC News*, October 18, 2018.
23. "Pallone & Markey Introduce Bills to Stop Robocalls." June 7, 2018 press release from Congressman Frank Pallone, Jr. The press release also cites potential profits for robocallers as much as $20 for every dollar spent.
24. "Just $6,790 of $208 Million in Robocall Fines Have Been Collected by the FCC," *techdirt*, April 5, 2019.
25. 2016 Update of Occurrence Rates for Offshore Oil Spills, July 13, 2016, submitted by ABS Consulting, Inc. to the Bureau of Ocean Energy Management.

7

Optimization and Decision Making

Learning Objectives

- Understand when and where optimization can be appropriate and useful.
- Appreciate how Excel Tables can be used to find optimal solutions for relatively simple problems.
- Learn how to use RiskOptimizer to find optimal solutions.
- Learn how to combine optimization and simulation.
- Understand the difference between Excel's Solver and RiskOptimizer.
- Understand potential trade-offs between risk and return.
- Learn how to construct and use an Efficient Frontier.

7.1 Introductory Case: Airline Seat Pricing

The airline industry has led the revolution in revenue management pricing.[1] On the whole, airfares are a complex set of fare classes, where each fare class is subject to a myriad of restrictions. Deriving the best set of fare classes to offer for a single route alone is highly complex, but the complexity increases dramatically as the number of route segments grows. The best price for a seat on any particular flight leg typically also depends on the observed and predicted demand for that seat by all potential passengers that might use that leg as part of their journey.

One of the first, and simplest, solutions to the single-leg pricing problem was proposed by Littlewood.[2] A quick search for academic articles using the terms "Littlewood," "pricing," and "revenue" shows 218 published articles, with 197 of them since the year 2000.[3] Littlewood considered a single flight leg with two fare classes. The demand for the lower fare class is assumed to arrive before the demand for the higher fare class. Littlewood's suggested solution to this pricing problem is basically to accept demand for the lower price class seats until the known definite revenue from selling one more low-price seat is exceeded by the expected revenue (which is uncertain) of selling the same seat at the higher price. We will examine whether this principle does indeed provide optimal pricing when the demand for seats is uncertain.

7.2 A Simulation Model of the Airline Pricing Problem

Consider a Wednesday morning flight from Philadelphia (in Pennsylvania, USA) to Albany (in New York, USA). The aircraft has 100 seats and two fare classes: Y (business class) that sells for $300 and Q (economy class) that sells for $100.[4] The demand for Y class seats is assumed to be distributed according to a Normal distribution, with a mean of 30 seats and a standard deviation of 10.[5] Q class demand is also normally distributed, but with a mean of 90 and a standard deviation of 40. In this example, we make the simplifying assumption that the Q class passengers make reservations before the Y class passengers. Thus, the decision problem becomes: How many Y class passengers to accept before closing the plane to Q class passengers? If we leave seats empty, there is the potential that there will not be any Y class passengers to demand them, but if we fill the seats with Q class passengers, then we may not have seats for the high-paying Y class passengers. Also, assume that any unmet demand due to lack of seats will be lost to other airlines.

Following Boyd's (2007) paper, we examine three strategies:

- Strategy 1: No control—Passengers are permitted to purchase tickets as long as seats are available.

- Strategy 2: Mean control—Since we expect an average of 30 Y class passengers, restrict Q class purchases to no more than 70. Clearly, this is the optimal strategy if the demand levels are certain (which they are not) and equal to their means (then 30 Y class seats and 70 Q class seats will always be sold).

- Strategy 3: Littlewood's rule—Restrict Q class purchases so that the *expected* incremental revenue for each fare class is equalized. Our initial spreadsheet will examine this as restricting Q class purchases to no more than 63 seats (this is what appears in Boyd [2007]), but we will see that it is not a correct statement of Littlewood's rule.

We will simulate revenue for Wednesday flights from these three strategies. The model (Airline Pricing7a.xlsx) first simulates the demand for Q and Y seats, and then calculates the revenues for the three strategies. Figure 7.1 shows the first few weeks and first two strategies of our initial model.

	A	B	C	D	E	F
1	Seats	100		price	mean	std dev
2			Y class	300	30	10
3			Q class	100	90	40
4						
5	Q demand	Y demand				
6	=RiskNormal($E$3,$F$3,RiskTruncate(0,))	=RiskNormal($E$2,$F$2,RiskTruncate(0,))				
7						
8						
9		Reserved Seats	Q sales	Y sales	Revenue	Empty Seats
10	Strategy 1	0	=ROUND(IF(A6<=$B$1,A6,$B$1),0)	=ROUND(IF(B6<$B$1-C10,B6,$B$1-C10),0)	=RiskOutput()+C10*$D$3+D10*$D$2	=RiskOutput()+$B$1-C10-D10
11						
12	Strategy 2	30	=ROUND(IF(A6<=$B$1-$B$12,A6,$B$1-$B$12),0)	=ROUND(IF(B6<$B$1-C12,B6,$B$1-C12),0)	=RiskOutput()+C12*$D$3+D12*$D$2	=RiskOutput()+$B$1-C12-D12
13						
14	Strategy 3	37	=ROUND(IF(A6<=$B$1-$B$14,A6,$B$1-$B$14),0)	=ROUND(IF(B6<$B$1-C14,B6,$B$1-C14),0)	=RiskOutput()+C14*$D$3+D14*$D$2	=RiskOutput()+$B$1-C14-D14

FIGURE 7.1
Initial Airline Pricing Model

A6 and B6 simulate the weekly demand for Q seats and Y seats. Note that the Normal distribution functions have an argument RiskTruncate(0,). To include this truncation functionality within @Risk, when entering these distributions, select Alternate Parameters at the drop-down menu shown in Figure 7.2, and enter a minimum truncation limit of 0: this is how we ensure that demand cannot be negative.

For each strategy, the Q sales are determined by comparing the demand with the seats available for Q class and using whichever is smaller.[6] Y sales are then constrained by the available seats minus those that have been sold to Q class passengers (which is why Strategy 1, no control, often shows no Y sales).

Given the two prices and two demands, Airline Pricing7a.xlsx examines the three pricing strategies described above: (1) no reserved seats (saved for Q class customers), (2) 30 reserved seats, and (3) 37 reserved seats. Figure 7.3 shows a single simulation iteration for the completed model.

In this particular iteration, the randomly generated demand for Q and Y class seats is 108 and 36, respectively. In this case, all 100 seats will go to Q class customers when no seat control is used, but both control strategies will limit Q class sales. Given that 36 Y class seats are demanded, Strategy 3 permits all these seats to be sold, while Strategy 2 only has 30 Y class seats available for sale.

Once we simulate the number of Q and Y class sales, it is straightforward to calculate the total revenue for the flight. We also keep track of how many empty seats are on each flight and mark those totals as outputs. Running the model and overlaying the total revenue boxplots gives Figure 7.4.

The no control strategy (Strategy 1) is clearly not the approach that results in the greatest revenues, but the mean control strategy and Littlewood's rule strategy appear to produce similar mean results. The mean revenue for Littlewood's rule (Strategy 3) is slightly greater than for the mean control strategy (Strategy 2), although it appears to be a bit more variable.[7] The results for the number of empty seats under each strategy are shown in Figure 7.5.

Strategy 3 has more empty seats than Strategy 2, but earns slightly higher revenues. This is due to the increased number of high-revenue Y class passengers that it allows. But, the question is whether Littlewood's rule produces the *best* result under the assumptions of this model. That is, is the cutoff of 63 Q class seats (37 Y class seats) the best cutoff under these conditions?

7.3 A Simulation Table to Explore Pricing Strategies

A simulation table is a possible way to address this question. Figure 7.6 shows the setup of the simulation table (now included in the model file Airline pricing7b.xlsx, where we have reduced the model to a strategy where reserved Y seats can vary between 30 and 40.

Running the simulations (remember to change the @Risk settings to run 11 separate simulations, each with 10,000 iterations) provides the results shown in Figure 7.7.

The highest mean revenue (shaded) occurs, not at 37 reserved seats, but at 34. We also see another finding: the standard deviation is lower when we use mean control (in fact, the standard deviation rises continuously as we reserve more seats for Y class passengers). So, the additional revenue that Littlewood's rule provides comes with a disadvantage—it increased the risk in the revenues *per flight*. In the present case, an airline with thousands of flights per day will not care much about this increased risk per flight, since it will cancel out across these many flights (central limit theory).[8] The expected revenues would be the more important measure for choosing a strategy, and that is why Littlewood's rule is stated in terms of the "expected revenue."

In fact, the application of Littlewood's rule would require the *expected* Y class revenue to equal $100 (the price of a Q class seat).

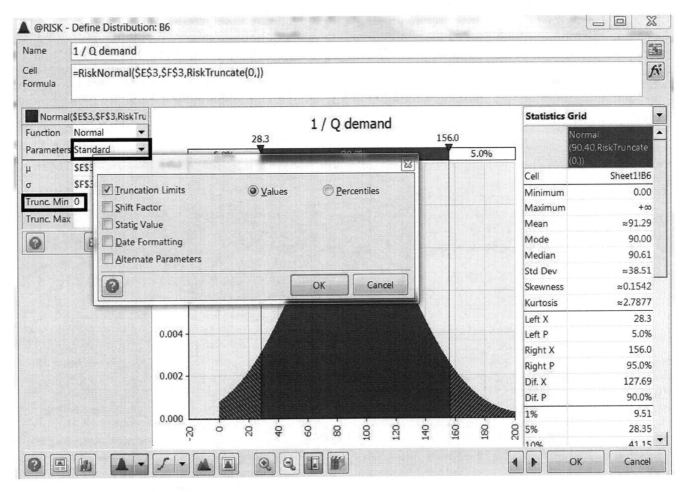

FIGURE 7.2
Entering Truncation in a Distribution

Optimization and Decision Making

	A	B	C	D	E	F
1	Seats	100		price	mean	std dev
2			Y class	$300	30	10
3			Q class	$100	90	40
4						
5	Q demand	Y demand				
6	109	28				
7						
8						
9		Reserved Seats	Q sales	Y sales	Revenue	Empty Seats
10	Strategy 1	0	100	0	$10,000	0
11						
12	Strategy 2	30	70	28	$15,400	2
13						
14	Strategy 3	37	63	28	$14,700	9

FIGURE 7.3
One Iteration of the Airline Pricing Model

Since the Y class price is three times the Q class price, this is equivalent to a finding where the probability of getting an additional Y class customer decreases to 33.3% or less. In other words, we should increase the seats reserved for first class passengers until the probability of getting more first-class travelers drops to 33.3% (thereby equating the expected revenue from an additional first-class passenger with the economy class revenue. Using the NORMINV(0.667,30,10) function in Excel, this is found to be 34.4. Our spreadsheet model permits us to discover Littlewood's rule as the revenue maximizing strategy.[9] It also shows us that maximizing revenue may not always be the best strategy, since a revenue (or profit)-maximizing strategy can come at the expense of exposing us to higher risk. In fact, what we often find with optimization models is that they can help us identify the strategy that best balances maximizing a certain objective (e.g., profit, revenues) with an acceptable amount of risk.

7.4 A Stochastic Optimization Solution to the Airline Pricing Problem

As we saw, using a simulation table within @Risk can be sufficient for finding the optimal number of Y class seats for a relatively simple problem where the number of possible decisions is limited. This approach has the additional benefit of showing exactly how revenues vary across the range of available decision choices. In more complex problems where the number of possible decision alternatives is (much) larger, it can, however, be desirable to automate this analysis—optimization refers to finding the best solution to such problem. When discussing a quantitative optimization, there are three important components to define when setting up the model. First, *decision variables* are quantities or values over which we have control. For example, when optimizing the allocation of money among a portfolio

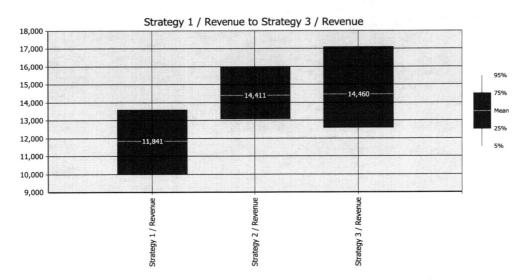

FIGURE 7.4
Initial Airline Pricing Results

of possible investment options, the decision variables include the percent of the portfolio allocated to each type of investment (e.g., stocks, bonds, real estate). For our airline pricing problem, our only decision variable is the number of seats to reserve for potential Y class customers.

Second, *constraints* are restrictions that "constrain" the values taken on by the decision variables. In a simple portfolio example, one possible constraint is that the sum of the allocations must be equal to 100%. In our airline pricing problem, a key constraint is that the available Q class seats plus the reserved Y class seats equal 100. In a project portfolio optimization, a constraint may be the available budget, or the number of FTEs.

Third, the *objective function* of the problem is the value or quantity we are trying to optimize, for example, the maximum expected return from our investment portfolio or the maximum total expected flight revenue. In other optimizations, we might be trying to minimize a value, such as trying to minimize the amount of inventory to keep in stock (e.g., while still meeting a certain fill). We often have more than one value we are trying to optimize, in which case we can optimize multiple objective functions, although this can technically be more difficult. In practice, most optimization problems are defined as a set of decision variables that we are trying to manipulate so as to maximize (or minimize) an objective function, taking into account a number of constraints.

Optimization and Decision Making

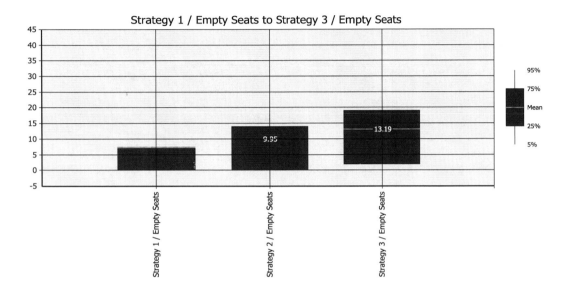

FIGURE 7.5
Empty Seats Results

There are many techniques available for solving optimization problems. We examine here a small subset of possible techniques. Trial and error (such as through a simulation table as shown above) is one of the most basic optimization methods and often works very well if there are a relatively small number of decision variable values to use. However, when there are many possible values for each decision variable (as with situations where there are several continuous decision variables) or too many potential combinations of decision variables (as with several or many decision variables, each with a number of possible values), then manual trial and error quickly becomes impractical or even infeasible. For example, if we had a simple situation with only 10 decision variables, where each could take on two possible values (0 or 1), there are already 2^{10} (1024) possible combinations. Depending on the specific problem, we might be able to try all 1024 combinations to determine the best values for the decision variables.

It is easy to see that for realistic problems, with many more decision variables, the number of possible outcomes can become astronomically large. To illustrate this, consider a problem with 50 binary decision variables (which is still smaller than most real-world scenarios). There are now 2^{50} (1,125,899,906,842,620) possible combinations. If we could check 1 million combinations per second

	H	I	J	K	L	M	N
1	=RiskSimtable(J2:J12)	Sim #	Reserved	Mean Revenue	Std Dev Rev	Mean empty	Std dev empty
2		1	30	=RiskMean($F$6,I2)	=RiskStdDev($F$6,I2)	=RiskMean($G$6,I2)	=RiskStdDev($G$6,I2)
3		2	31	=RiskMean($F$6,I3)	=RiskStdDev($F$6,I3)	=RiskMean($G$6,I3)	=RiskStdDev($G$6,I3)
4		3	32	=RiskMean($F$6,I4)	=RiskStdDev($F$6,I4)	=RiskMean($G$6,I4)	=RiskStdDev($G$6,I4)
5		4	33	=RiskMean($F$6,I5)	=RiskStdDev($F$6,I5)	=RiskMean($G$6,I5)	=RiskStdDev($G$6,I5)
6		5	34	=RiskMean($F$6,I6)	=RiskStdDev($F$6,I6)	=RiskMean($G$6,I6)	=RiskStdDev($G$6,I6)
7		6	35	=RiskMean($F$6,I7)	=RiskStdDev($F$6,I7)	=RiskMean($G$6,I7)	=RiskStdDev($G$6,I7)
8		7	36	=RiskMean($F$6,I8)	=RiskStdDev($F$6,I8)	=RiskMean($G$6,I8)	=RiskStdDev($G$6,I8)
9		8	37	=RiskMean($F$6,I9)	=RiskStdDev($F$6,I9)	=RiskMean($G$6,I9)	=RiskStdDev($G$6,I9)
10		9	38	=RiskMean($F$6,I10)	=RiskStdDev($F$6,I10)	=RiskMean($G$6,I10)	=RiskStdDev($G$6,I10)
11		10	39	=RiskMean($F$6,I11)	=RiskStdDev($F$6,I11)	=RiskMean($G$6,I11)	=RiskStdDev($G$6,I11)
12		11	40	=RiskMean($F$6,I12)	=RiskStdDev($F$6,I12)	=RiskMean($G$6,I12)	=RiskStdDev($G$6,I12)

FIGURE 7.6
Simulation Table for Reserved Seats

(which is unreasonably fast for a spreadsheet model) it would still take over 35 years to check every combination in order to determine an optimal solution. Fortunately, there are many optimization software programs available that can automate, and speed up, the process of solving optimization problems.

To conceptually understand how an optimization solver might work, consider a simple objective function and a single decision variable, as shown in Figure 7.8.

Suppose a value for the decision variable, such as D1, is evaluated with a model and produces the result for the objective function, V1. Then, slightly higher and lower values than D1 (as indicated by the arrows) are tried, and the decision variable is moved in the direction (in this case, increasing D) that produces higher values for the objective function. If a value of D > D* is tried, then the objective function will increase by reducing the value of D. Eventually, after trying a number of different values of D, D* (or a value very close to D*) will be found to solve this optimization problem. This technique, called a "hill-climbing" technique, is straightforward—provided that the objective function is well-behaved, as in Figure 7.8. However, what if the objective function takes the form shown in Figure 7.9?

Depending on the starting value, a simple hill-climbing method could easily decide that D1 or D3 is the optimal decision, whereas D2 is clearly superior. When there are multiple local maxima in the objective function and/or discontinuous segments (such as would be produced when there are IF statements in the spreadsheet), then a smarter procedure is required. Such smarter optimization methods generally employ techniques aimed at finding the true global optimum, even in the face of such "discontinuity" features.

These techniques differ in optimization methods and programs, but when considering spreadsheet-based simulation models such as we have been working with, the optimization methods within @Risk attempt to mimic biology in finding a solution. Such evolutionary

Optimization and Decision Making

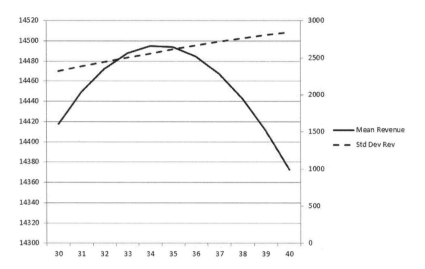

FIGURE 7.7
Simulation Results for Varying Reserved Seats

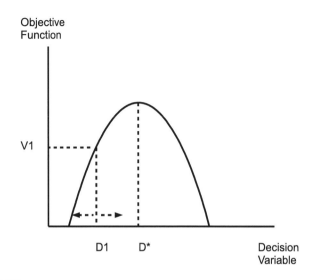

FIGURE 7.8
Simple Optimization Problem

or *genetic algorithms* employ random jumps in the decision variable (analogous to mutations) to prevent premature convergence to a local, and potentially suboptimal, solution. By ensuring that more potential regions of the decision variable are explored, these techniques can be effective at finding a good (or even the global) optimum, even when the objective function is not well-behaved (i.e., not smooth/continuous).

Excel comes with an optimization procedure built in, called Solver. It is found under the Data ribbon of the Excel toolbar.[10] The basic Solver that comes standard with Excel is limited in the type of optimization problems it can solve—in particular, it is not always reliable when there are conditional statements (such as IF, Max, Lookup statements) in the model. Solver also frequently may get stuck in a local optimum. Excel's free Solver can handle the airfare pricing model without simulation, but it cannot accommodate including Monte Carlo simulation within the optimization.

Fortunately, alternative and more powerful optimization methods and tools are available.[11] We will use RiskOptimizer, which is an optimization tool included with @Risk, found on the @Risk toolbar. RiskOptimizer is designed to be well suited for solving optimization models that also contain Monte Carlo simulation. RiskOptimizer imposes no restrictions on the mathematical form of the optimization model, or the shape of the objective function, so it can be used in our airfare pricing problem. RiskOptimizer will be able to find the decision variables that maximize the objective, but it can do so while

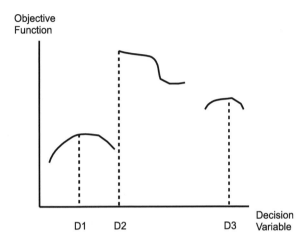

FIGURE 7.9
A More Complex Objective Function

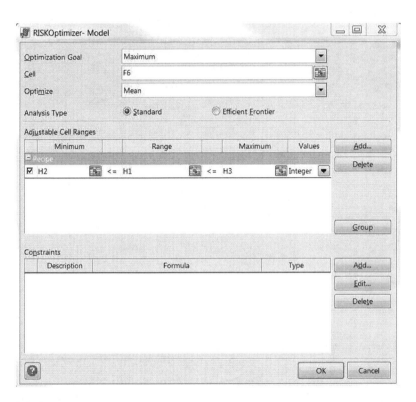

FIGURE 7.10
Completed RiskOptimizer Dialog

taking into account risk by conducting a simulation for each possible scenario. RiskOptimizer is capable of solving either deterministic or stochastic (i.e., with or without Monte Carlo simulation) optimization.

To run RiskOptimizer for the airline pricing problem, we made a few small adjustments to the spreadsheet, as shown in the file Airline Pricing7c.xlsx. Choosing RiskOptimizer, Model Definition, we also fill in the dialog as shown in Figure 7.10.

We set the Optimizer to maximize the expected total revenue cell (notice that a number of alternative options to the "mean" are available). We have a single decision variable in cell H1, the number of reserved Y seats. Decision variables can be discrete or continuous,[12] and multiple decision variables can also be specified.

We will permit the number of reserved seats to range from 25 to 50 and only allow integer values. The next step, which deals with decision constraints, is optional (i.e., not each optimization model has/needs constraints), and in the current model, no constraints are necessary. A common constraint would be to constrain the decision variables to be nonnegative (but this is unnecessary if the range is a bounded set between two positive values like we have here). We could constrain the total seats sold to be less than or equal to 100, but that is not necessary since we have already constructed the model so that it automatically does not exceed the total capacity of the plane.[13]

There is a Settings option under RiskOptimizer that should be considered before running an optimization. In the current simple model, no adjustments are required to the default settings, but in a more complex model, it would be desirable to limit the time, number of iterations, or perhaps the convergence criterion (maximum change at which to stop). The settings and choices regarding the Efficient Frontier functionality will be considered later in this chapter.

Once the setup is complete, click "OK" and then choose "Start" from the RiskOptimizer menu button and obtain the results shown in Figure 7.11. There is a Progress Window at the bottom left of the screen and a button that will bring up the Watcher, where you can watch the progress of the optimization. Note that we decreased the number of iterations to 5000[14] since RiskOptimizer will be running a large number of trials (i.e., alternative values for the decision variables) and thus setting the number of iterations too high will cause the optimization to take a very long time.

We see that RiskOptimizer was indeed able to find the correct solution to reserve 34 seats for Y class customers. Figure 7.11 shows the summary results. The Log will show all the decision values RiskOptimizer used, along with the results of those values.[15]

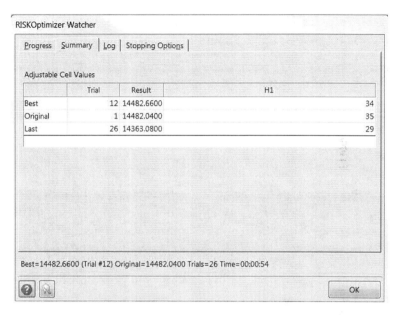

FIGURE 7.11
Airline Pricing Optimization Results

7.5 Optimization with Multiple Decision Variables

A real-life airline pricing problem is much more complicated in several ways. The problem we have thus far discussed considers only one route at a time, in a complex network of routes that airlines need to consider simultaneously. Further, the airline could set different (i.e., more optimal) prices for the two fare classes (not necessarily the $100 and $300 prices), could establish additional fare classes, and the ticket prices could change over time. In this section, we tackle some more of the complications associated with setting prices as well as reserved seats for Y class passengers. It is not sufficient to just designate prices as decision variables—to do so, we would discover that the higher the prices, the higher the revenue. In order to understand the trade-offs with higher prices, we must also alter the structure of the model so that raising the prices (for either class) will negatively affect the demand for seats.

Assume that every $10 increase in the Q class price will, on average, lead to a decrease of 9 in the expected demand (from a starting value of 200). Also assume that each $10 increase in the Y class price leads to a decrease of 1 in the expected Y class demand (from a base of 60).[16] Furthermore, assume that both mean demand functions are linear, that is,

(Q demand) = 200 − 0.9*(Q price) and (Y demand) = 60 − 0.1*(Y price)

We also assume that the standard deviations of both demand functions are increasing functions of price, that is, standard deviation = 7 + 0.01 * price. The spreadsheet model that takes these assumptions into account (Airline pricing7e.xlsx) is shown in Figure 7.12.

In this updated model, the expected (mean) demands (E2:E3) and standard deviations (F2:F3) are now expressed as a function of the prices (D2:D3), and these cells are used in the Normal distributions (B6:C6).[17] Note that the truncation of the demand distributions to prevent negative values becomes more important now. The variability in the simulations could easily produce negative demands if we did not truncate the distributions. The last required change is on the Decision Variables setting, where we need to add the two prices as two additional decision variables. In this area, we could also restrict some the ranges for the potential prices. Given our previous assumptions on price elasticity, Q class prices above 222 would generate negative demand, as would Y class prices above 600. Even though we have not permitted demand to be negative, it is not logical to consider prices in these ranges. In addition, negative prices do not make sense. So, we restrict prices to be within these ranges (i.e., if we had not added those ranges, in this case the model would had the same optimal solution, but it would likely have taken longer to find it). For our initial simulation, make the prices discrete with fairly large steps of $10–$20—we want to get a sense of what the optimal solution might look like before trying to get more precise. We also consider reserved seats between 25 and 50 with large steps (equal to 5), since we really don't know what the optimal solution might look like for the new problem. Figure 7.13 shows the Decision Variables settings.

We stop RiskOptimizer after 5 minutes, but it has not tried all the decision variable combinations. The total number combinations is 1800 (6 values of reserved seats * 20 Y class prices * 15 Q class prices), but RiskOptimizer has only run 315 trials at this point. The simulation summary for the progress steps allows us to focus the range of the decision variables considerably. Figure 7.14 shows these progress steps.

FIGURE 7.12
Multiple Decision Variable Airline Pricing Model

Optimization and Decision Making

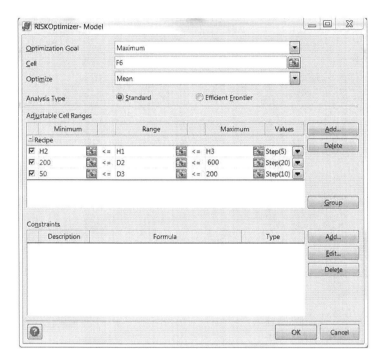

FIGURE 7.13
Initial Settings for Optimization with Multiple Decision Variables

Based on these fairly narrow ranges for the top solutions, we will rerun RiskOptimizer with the narrower ranges and smaller increments shown in Figure 7.15.

With the narrower ranges for the decision variables, the optimization (after 5 minutes, having run 500 trials out of 605 possible combinations) produces the solution analysis shown in Figure 7.16.

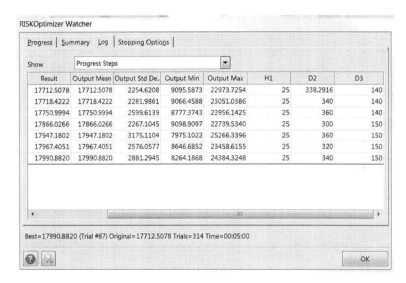

FIGURE 7.14
Progress Steps for the Initial Optimization

At this point, it looks like the best solution is to reserve 28 seats for Y class, with a price of 335, and sell Q class tickets at a price of 150. It is worth noting that the best solution, without accounting for the uncertainty in the demand, is to set the Y class price at 330 and Q class price at 141.[18] Using the best RiskOptimizer solution at this point shows an average of 10.9 empty seats on the flight. Taking into account that demand is uncertain therefore results in an optimal solution that recommends reserving a few more high-priced Y seats than under certainty, and charging slightly higher Y class prices. We leave it to the reader to explain (in layman terms) why this difference in the two solutions makes sense.

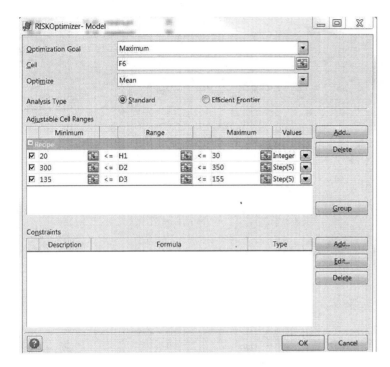

FIGURE 7.15
More Precise Optimization Settings

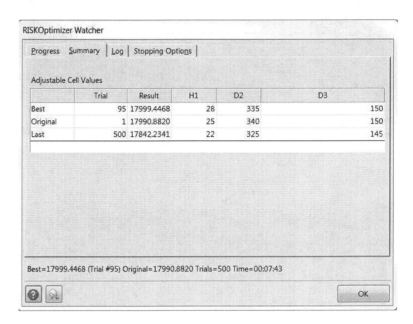

FIGURE 7.16
More Precise Optimization Solution for Three Decision Variables

7.6 Adding Constraints

Many optimization problems are *constrained optimization* problems. Constraints can take the form of direct constraints on decision variables (e.g., that prices are nonnegative, that total portfolio asset allocations must add up to 100%, etc.), indirect effects on the effects of decision variables (e.g., that demand that is based on prices be nonnegative, that no portfolio share exceed 25%, etc.), or constraints on the target variable we are optimizing itself (we will examine this type of constraint in the next section).

Some direct constraints on decision variables can easily be incorporated into the model or the ranges set in RiskOptimizer (such as not permitting negative prices). The example we provide here cannot easily be built directly into the model. The optimal solution for our three decision variable problem from the last section was $P_Y = \$335$ and $P_Q = \$150$ ($330 and $141 are the true

Optimization and Decision Making

optimal values we would obtain if we ran RiskOptimizer long enough with greater granularity). However, the airline may be concerned about business class customers flying economy class if the price difference is too large. Suppose that the airline therefore does not want business class prices to be more than twice the level of economy class prices. We can model this by adding a constraint to our optimization problem. Figure 7.17 shows the constraint setup after clicking Add to the constraint box of the optimization model setup (where we have added a constraint formula in cell D4 = 2 * PQ in Airline Pricing7e.xlsx).

If you observe the RiskOptimizer watcher during the simulation, you will see that it begins (from the prior solution) with N/A, indicating that the original solution does not satisfy the constraint. However, fairly quickly, RiskOptimizer find a solution (from the current decision variable parameters) of $P_Y = \$310$ and $P_Q = \$155$. While a more exact solution may be warranted (by making the optimization search more granular in both this section and the prior one), the additional constraint has led to a slightly lower business class price and slightly higher economy class price. With both prices suboptimal (compared with an unconstrained problem), the expected revenue of the optimal situation is now $17,898 rather than the $17,999 found in Section 7.6. In other words, this constraint "costs us" $101 in expected revenue.

Incorporating constraints permits us to understand the optimal solution when there are (additional) constraints, as well as providing information about the costs these constraints impose on our objective variable (e.g., the revenues).

7.7 Efficient Frontier

We return to our unconstrained multiple decision variable problem from Section 7.6. The optimal solution maximizes the expected revenue, but does result in variability around the expected revenues (the expected revenues are $17,999 and the standard deviation is $2845). We might consider this standard deviation too large—in other words, we might find the solution with the maximum expected revenues is too "risky." Alternatively, running the simulation for the optimal prices reveals that the probability that revenues will be less than $14,000 is 10.7%. We might deem this too high and might only tolerate that probability to be 7% or less. Incorporating constraints such as these is accomplished by imposing constraints on the objective (target) variable.

For example, Figure 7.18 shows the Constraint Settings just described.

Note that the radio button for our constraint is changed from "Value" to "Simulation Statistic" to accommodate this type of simulation constraint. We are now showing the results of the optimization model with this additional constraint. Instead, we will go one step further and examine the *sensitivity* of the optimal objective variable to a range of measures of its variability. These risk-return trade-offs can often be examined with something called an *efficient frontier*. This term is widely used in finance, but it can

FIGURE 7.17
Optimization Constraint Setup

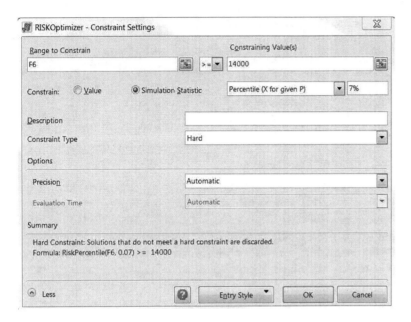

FIGURE 7.18
Constraint Settings on Variability of an Objective Variable

refer to any relationship that shows the optimal value of an objective function as a function of various levels of a simulation constraint (such as the risk). Construction of an efficient frontier entails running the optimization at a number of values for the simulation constraint, collecting the results, and then displaying the frontier: for example, the maximum return for each level of risk.

To derive an efficient frontier, change the radio button on the Optimization Model from "Standard" to "Efficient Frontier" and add the constraint as shown in Figure 7.19 (and is contained in Airline Pricing7f.xlsx).

FIGURE 7.19
Constraint Settings for Efficient Frontier

As mentioned, our unconstrained solution in Section 7.6 entailed a standard deviation of total revenues of $2845. So, the upper end of our constraint range will allow for that amount of risk for the

Optimization and Decision Making

optimal solution. The steps with lower standard deviations will imply constraints on our optimization problem with increasingly low risk.

Before running the model, we expand the ranges for our decision variables. With these new constraints, we do not yet know what values of the decision variables will be feasible solutions. Figure 7.20 shows our new model.

We could have used even larger ranges for our decision variables and less granular steps—generally this is advisable, as optimization is best viewed as an iterative search for a solution. Alternatively, if we find that Risk Optimizer finds a solution at the boundary of where we allowed it to search (or fails to find a feasible solution), that is an indication that we need to enlarge the range of potential decision variable values.

Running the simulation and opening the Watcher shows the construction of the frontier as the constraint is varied. For this example, we stop the optimization after 10 minutes. The normal procedure would be to iteratively run Risk Optimizer with increasingly granular ranges for decision variables and constraints, and run for a sufficient length of time to ensure we are close enough to the optimal solution. Figure 7.21 shows a snapshot of the construction of the frontier at this point.

Note that the optimal values are not increasing very much during the last few hundred trials—one indication that we may be getting close to the optimal solutions (although more granularity will still be desired in most applications). After 10 minutes, Risk Optimizer has found feasible solutions for all the constraint values, and you can generate a number of reports for the simulations. Figure 7.22 shows the Summary report with the Efficient Frontier.

Note that none of the decision variable solutions (the "Adjustable Cells") fall on the boundaries of the ranges we limited Risk Optimizer to search over. It appears that the less variable (i.e., risky) we constrain the revenues to be, the lower the business class price we need to set. Economy prices, on the other hand, do not vary much,

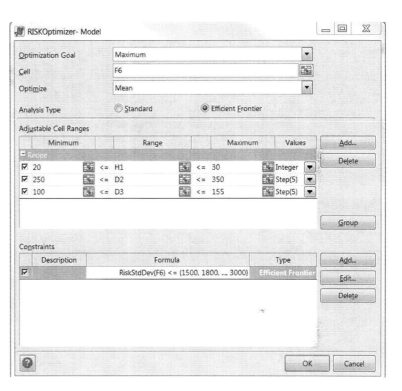

FIGURE 7.20
Efficient Frontier Model Setup

nor do the Y class reserved seats. The Efficient Frontier provides a visual view of the trade-off between risk (the horizontal axis) and return (the vertical axis). Such insights into which decision variables mostly drive the trade-off between expected revenues and risk can now help management in deciding on a strategy that best fits their risk tolerance.

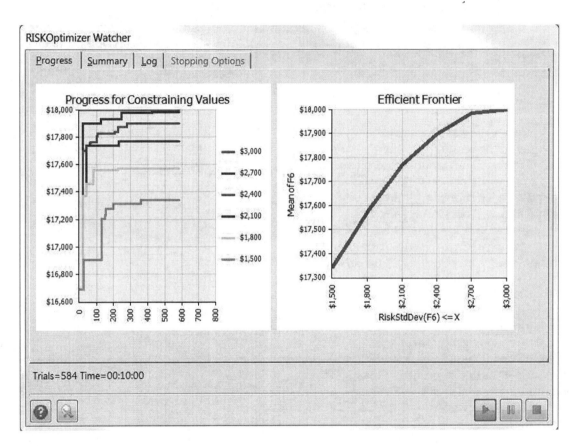

FIGURE 7.21
Constructing the Efficient Frontier

Optimization and Decision Making

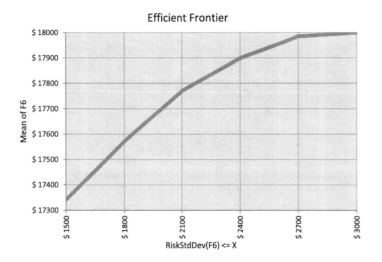

Constraining Value	Valid Trials (for constraining value)	Best Value	Trial	Goal Cell Statistics				Adjustable Cells			Hard Constraints
				Mean	Std. Dev.	Min.	Max.	H1	D2	D3	RiskStdDev(F6)
$1,500	219	$17,341	343	$17,341	$1,492	$10,058	$20,769	24	$270	$145	$1,492
$1,800	297	$17,571	227	$17,571	$1,766	$9,888	$21,232	30	$280	$145	$1,766
$2,100	384	$17,770	244	$17,770	$2,049	$9,523	$22,230	29	$300	$145	$2,049
$2,400	452	$17,899	268	$17,899	$2,344	$8,982	$22,888	25	$305	$150	$2,344
$2,700	538	$17,985	424	$17,985	$2,663	$8,543	$23,707	26	$325	$150	$2,663
$3,000	564	$17,999	1	$17,999	$2,845	$8,351	$24,205	28	$335	$150	$2,845

FIGURE 7.22
Efficient Frontier Results

7.8 Stochastic Dominance

In some situations, when comparing[19] distributions for two different scenarios (e.g., the NPV of project A and project B), it will not be immediately obvious which one is better. It may be easy to see which has a higher expected NPV, but often a project that results in higher expected value will also be a project that has greater risk. There is usually a trade-off involved between risk and expected return.

Stochastic dominance is a technique that allows you to compare two (or more) output distributions and determine whether one is "better" (higher or lower) than the other. When using a stochastic dominance test, the most common comparisons involve determining whether one distribution stochastically dominates another in the first degree or the second degree. First-degree stochastic dominance means that one cumulative distribution is always to the right of the other. Consider the distributions for the profitability or NPV of two alternative projects shown in Figure 7.23.

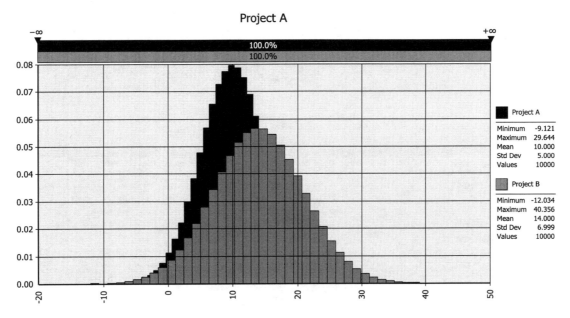

FIGURE 7.23
Two Overlapping Distributions

Optimization and Decision Making

It is not immediately obvious which project is to be preferred. Project B appears to have a higher expected return, but also appears to be more variable. However, if we view the cumulative distribution functions, the choice becomes clear. Figure 7.24 shows the cumulative probability distributions.

In Figure 7.24, Project B's cumulative distribution of the objective function is always greater than that for Project A. This means that there is always a greater probability that Project B will reach any given level of profitability than will Project A. Clearly, any decision maker should prefer Project B to Project A.

A distribution dominates another in the second degree when any risk-averse decision maker would prefer one to the other. An example would be two distributions with the same mean, but one with a higher standard deviation than the other, as shown in Figure 7.25.

In cases like this, any risk-averse decision maker would prefer the narrower distribution since it has less uncertainty, but the same expected payoff. In general, second-degree stochastic dominance is not very "selective" and not often used since typically other reasons (beyond the quantitative results) can be thought of to prefer one option versus another.

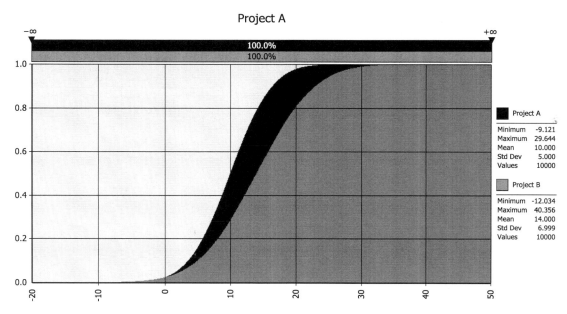

FIGURE 7.24
Cumulative Distributions for Two Projects

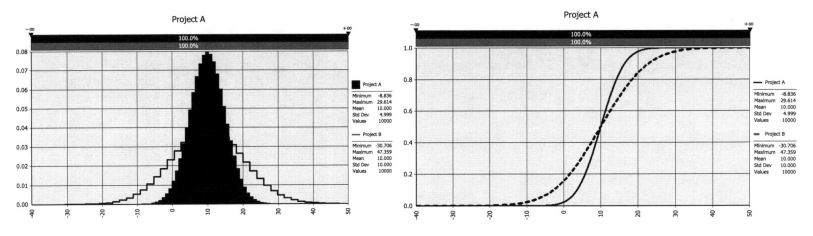

FIGURE 7.25
Example of Second-Degree Stochastic Dominance

Now, can stochastic dominance be applied to our airline pricing problem? In Section 7.2, we produced a simulation table that revealed that reserving 34 seats for Y class produced the highest expected revenue, but that reserving only 30 seats produced a lowest standard deviation. Can we compare these two strategies using stochastic dominance?

Airline pricing7g.xlsx is the original spreadsheet (Airline pricing7a.xlsx), adapted to consider stochastic dominance between the two strategies: reserve 30 seats and reserve 40 seats (the latter a strategy with greater risk). The two strategies (30 and 40 reserved seats) have been inserted into the Strategy 2 and 3 rows, respectively. To use the stochastic dominance test, we need to compare the output cumulative distributions for the two strategies. We run the simulation and then extract the simulated percentiles for the two cumulative distributions (worksheet "Dominance") for the two outputs by copying the data grid from the @Risk results cumulative overlay and pasting this into the worksheet. Figure 7.26 shows what we are pasting.

Second-degree stochastic dominance depends on whether the area under the curve showing the *differences* between the two cumulative probability distributions is positive for all values of the percentiles. We compute the differences between these pasted cells (total revenue with 30 reserved seats minus total revenue with 40 reserved seats) and graph them as a function of the percentiles, as shown in Figure 7.27, along with an approximation[20] of the total area under the difference curve.

The cumulative *difference* of the area is positive (as visual inspection suggests). This means that the strategy of reserving 30 business class seats dominates the strategy of reserving 40 business class seats in the second degree.

Optimization and Decision Making

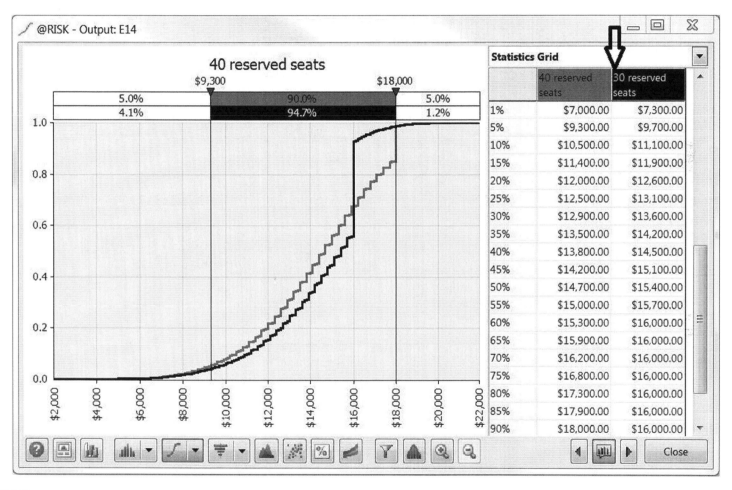

FIGURE 7.26
Percentiles for Stochastic Dominance

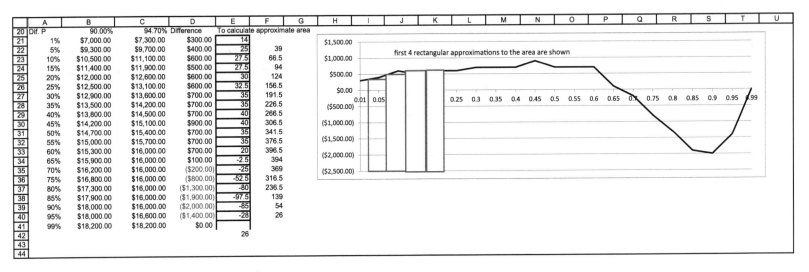

	A	B	C	D	E	F	G
20	Dif. P	90.00%	94.70%	Difference	To calculate approximate area		
21	1%	$7,000.00	$7,300.00	$300.00	14		
22	5%	$9,300.00	$9,700.00	$400.00	25	39	
23	10%	$10,500.00	$11,100.00	$600.00	27.5	66.5	
24	15%	$11,400.00	$11,900.00	$500.00	27.5	94	
25	20%	$12,000.00	$12,600.00	$600.00	30	124	
26	25%	$12,500.00	$13,100.00	$600.00	32.5	156.5	
27	30%	$12,900.00	$13,600.00	$700.00	35	191.5	
28	35%	$13,500.00	$14,200.00	$700.00	35	226.5	
29	40%	$13,800.00	$14,500.00	$700.00	40	266.5	
30	45%	$14,200.00	$15,100.00	$900.00	40	306.5	
31	50%	$14,700.00	$15,400.00	$700.00	35	341.5	
32	55%	$15,000.00	$15,700.00	$700.00	35	376.5	
33	60%	$15,300.00	$16,000.00	$700.00	20	396.5	
34	65%	$15,900.00	$16,000.00	$100.00	-2.5	394	
35	70%	$16,200.00	$16,000.00	($200.00)	-25	369	
36	75%	$16,800.00	$16,000.00	($800.00)	-52.5	316.5	
37	80%	$17,300.00	$16,000.00	($1,300.00)	-80	236.5	
38	85%	$17,900.00	$16,000.00	($1,900.00)	-97.5	139	
39	90%	$18,000.00	$16,000.00	($2,000.00)	-85	54	
40	95%	$18,000.00	$16,600.00	($1,400.00)	-28	26	
41	99%	$18,200.00	$18,200.00	$0.00		26	

FIGURE 7.27
Difference in Cumulative Distributions

Stochastic dominance tests can provide a measure with which to compare the performance of different strategies, where each strategy has different return and risk levels.

7.9 Summary

There is a joke about a traveler landing in New Zealand and renting a car. The traveler knows that New Zealand was part of the British Empire but can't recall whether they drive on the left or right side of the road. So, the traveler decides to drive in the middle, with the reasoning that, on average, it is the correct choice.

Such "optimization," using the "average" of possible outcomes and ignoring the varying scenarios, may only hit the target of disaster. Used properly, stochastic optimization can greatly assist with decision making under uncertainty. In particular, when there are multiple decision variables or when attitudes toward uncertainty can be explicitly specified, then the use of optimization tools can be quite robust.

Simulation tables and graphical exploration of simulation results can always be considered as an alternative to, or complement to, optimization algorithms that are part of software-based packages like RiskOptimizer. Usually, different potential decisions will imply different risks and returns. Occasionally, the choices between alternative solutions will be easy, but most of the time they will involve complex trade-offs that are not always easy to understand and rank qualitatively. Analysts can support decision makers by elucidating these trade-offs. Optimization together with Monte Carlo simulation will allow us to identify the optimal solution while acknowledging and taking into account risk and uncertainty.

TABLE 7.1

Treatment Options

Treatment	Cost per Treatment	QALYs	Expected Lifetime	Year of Full Recovery
A	$131,200	8.26	10.2 years	5
B	$94,800	7.68	9.1 years	4
C	$76,200	6.48	8.7 years	never

The expected lifetimes reflect the average remaining life for people who contract this condition.

Exercises

Exercise 7.1 Quality-Adjusted Life Years

Medical decisions are complex, and many affect both the length of life and the quality of life. Quality-adjusted life years (QALYs) is a metric that has been developed to assist with such decisions.[21] The basic idea behind QALY is that a year of life in perfect health equals one QALY. If a person is disabled or in sufficient pain, a single year might be equivalent to only one-half quality adjusted year. More (less) severe impairments equate to smaller (larger) QALYs. Many medical treatments vary in terms of how they influence both the length and quality of life. For example, more invasive treatments often involve prolonging life, but at the expense of a low quality of life, at least during the period of recovery from the treatment. Treatments also vary in terms of cost.

Consider the hypothetical treatments described in Table 7.1. Assume the discount rate is 10%. Treatment A is fairly costly and invasive, and entails a long recovery period, but full recovery is expected. Treatment C is less costly and invasive, but never completely cures the patient, leaving some chronic problems. Treatment B is in between.

a. Build a model to calculate cost per present value of QALYs for each treatment. Use a 15-year time horizon and round fractional years downward to the nearest integer for calculating the benefits each year (e.g., 10.2 years is 10 years—do not use partial years for the timeline, but use the values in the table for calculating the QALYs).

b. Assume that the expected lifetimes are uncertain and follow Lognormal distributions (with means equal to those given in Table 7.1 and standard deviations equal to 30% of these values). The QALY data is also uncertain: these follow Lognormal distributions (with means given in Table 7.1 and standard deviations equal to 20% of these values). Build a simulation model to compare the cost per QALY (discounted). How often does each treatment minimize the cost per QALY?

c. Suppose you have $1 billion to spend on these treatments and there are 20,000 people that suffer from this condition. How would you allocate your budget among the three treatments to maximize the total discounted QALYs? How many people will be treated with this solution? How would you allocate your budget to maximize the number of people treated?

Exercise 7.2 Fuel Costs vs. Labor Costs

Suppose a truck gets 11 miles per gallon if it travels at an average speed of 50 miles per hour. Fuel efficiency is reduced by 0.15 mpg for each increase in the speed of 1 mph, up to 60 mph. Fuel efficiency is reduced by 0.2 mpg for each 1 mph in the range from 60 to 70 mpg, and by 0.3 mpg for each 1 mph above 70 mph. Drivers earn $50 per hour and fuel costs $3.20 per gallon.

a. What speed minimizes the total cost of a 500-mile trip? Does it depend on the length of the trip?

b. Suppose fuel costs are uncertain, following a Lognormal distribution (mean = $3.20/gallon, standard deviation = $1). What speed minimizes the total expected trip cost? Again, does it depend on the length of the trip?

c. With uncertain fuel costs (as specified in part [b]), what speed minimizes the total cost of a 500-mile trip with a 90% probability that the total cost will be less than $600?

Exercise 7.3 Portfolio Management

Exercise7-3.xlsx contains annual data for the inflation-adjusted returns on stocks, Treasury bills, and Treasury bonds from 1928–2018. Imagine that you have $100,000 to invest in these three funds and that you are currently 60 years old.

a. Build a simulation model to show the size of your portfolio over the next 5 years for an equal allocation among the three funds. Your model should be a time series model, including correlation. Choose an appropriate time period to use for your estimation and justify your choice. Run a simulation and produce a 90% confidence interval for the size of your fund after 5 years.

b. Find the portfolio allocation that maximizes your final value in 5 years. Provide 90% confidence intervals for your returns. Suppose the investor wishes to maximize the return but ensure that the probability of losing money is no greater than 10%. Construct an efficient frontier showing the optimal portfolio as the probability of losing money varies between 5% and 10% (in 1% increments).

c. Repeat part (b) for an investor that is 50 years old, planning for a 15-year investment period.

Exercise 7.4 Optimal Inventory Management

The simplest inventory model is the economic order quantity (EOQ) model, which assumes demand is known with certainty, there is a constant lead time between order and receipt of inventory, there is always a fixed order quantity, and there are constant holding and ordering costs per unit. More realistic inventory models are often based on this basic EOQ model as a starting point. A continuous review inventory system involves constant monitoring of the inventory on hand (and already ordered), and placing another order whenever the *inventory position* (current inventory + orders not yet received) falls below a level R, the *reorder point*. The general result in this model is that the optimal order quantity is found from Equation 7.1:

$$Q^* = \sqrt{\frac{2D(OC)}{HC}} \qquad (7.1)$$

where D is the annual demand, OC is the ordering cost per order, and HC is the holding cost per unit per year. Optimal inventory policy is a selection of the order quantity, Q^*, and the reorder point, R.

a. Consider a hot tub retailer with average annual demand for 620 tubs. Ordering is expensive, at $500 per order, as are inventory holding costs ($500 per tub per year). Assume that each order takes 3 weeks from the time it is ordered until it is delivered. If an order is received and there is no stock on hand, the customer is offered a rebate of $300 for having to wait for their tub. Build a model to show 52 weeks of evolution for this business. Find the optimal reorder point and order quantity for a year (that minimizes the total cost—remember that the reorder point and order quantity are single choices and do not change over time in this model). Given this optimal policy, what is the probability of not being able to make a sale that is demanded?

b. Suppose that demand is uncertain and follows a Poisson process with the average rate given above. Also assume that the delivery time (the time between ordering and receiving an order) is also uncertain: there is a 50% chance it will take 3 weeks, but a 25% chance it will take only 2 weeks, and a 25% chance it will take 4 weeks. Build a simulation model for 52 weeks and find the optimal order quantity and reorder point for this model. With the optimal policy, what is the probability of not being able to fulfill a retail demand when it arrives?

Exercise 7.5 Project Selection

Organizations often must make decisions regarding what R&D projects to pursue with limited resources available. Table 7.2

TABLE 7.2

Project Returns and Costs

Project	Expected NPV (Million CZK)	Expected Investment Cost (Million CZK)	Number of Workers
Project 1	22.3	31.0	23
Project 2	5.16	27.6	12
Project 3	28.21	126.8	45
Project 4	26.32	96.5	37
Project 5	15.32	55.8	25
Project 6	10.47	36.8	23
Project 7	12.30	44.7	15
Project 8	23.01	67.5	26
Project 9	15.07	49.0	24
Project 10	29.47	85.8	38
Project 11	20.24	53.3	20
Project 12	12.35	42.3	17

contains data on 12 hypothetical projects.[22] All monetary figures are measured in present values. Assume that both the net present values and investment costs are uncertain: assume that the distributions are lognormal, with means given by the figures in Table 7.2 and standard deviations equal to 50% of the mean values.

a. Pursuing all 12 projects would require 717.1 million CZK and 305 workers. Assume that a budget of only 560 CZK and a workforce of 240 are available. Find the optimal project portfolio (to maximize the expected NPV of the portfolio) with these constraints.

b. Suppose that the variability of this optimal portfolio is deemed unacceptably high. What is the optimal portfolio and its value if we wish to reduce the variability (measured by the standard deviation of the portfolio value) by 30%?

c. Produce efficient frontiers to show how larger budgets or larger workforces would affect the optimal expected portfolio value. Comment on the two constraints and which one appears to be a more severe constraint.

Exercise 7.6 Fisheries Management

Many models have been developed for the reproduction of fisheries and the effect of fish harvesting on fish populations. A very simplified version is provided here, calibrated for the Pacific Halibut Fishery.[23] This model is described by a few equations (where t refers to the discrete time period, e.g., a year):

Catch(t) = λ*Stock(t) where λ is the fraction of the stock that is harvested.

Escapement(t) = (1–λ)*Stock(t) where Escapement refers to the fish not harvested.

Reproduction(t) = Escapement(t)*Growth(Escapement(t)) where the Growth function is assumed to be a logistic function given by Growth = r*(1 – (Stock(t)/K) and r $\approx$ 0.4145 and K $\approx$ 551.6 (measured in millions of pounds) have been estimated for the Pacific Halibut Fishery.

a. Build a static model of the Pacific Halibut Fishery, without uncertainty, for 50 years, starting with the 2007 estimates of 136.344 for the stock of fish and 26.2 for the harvest rate (measured in millions of pounds). Assume that the harvest rate fraction of the stock remains constant over time.

b. Assume the beginning stock (2007) is uncertain (use a Pert distribution with a +/−20% range), but that the harvest rate fraction remains a constant fraction of the mean estimate of the beginning stock of halibut. Estimate the halibut population over the 50-year period, along with the probability of extinction.

c. It may be reasonable to expect that the price of halibut will rise as the harvest rate declines, and that one result might be more intensive fishing effort. Build this into your model by assuming that the fish catch fraction is not fixed, but instead assume that the fraction of fish caught increases by 0.01 for each 1 million decrease in the harvest rate. Redo part (b).

d. Estimate the maximum harvest rate obtainable at a 50-year time horizon (from 2007). Estimate the probability of extinction at 50 years if the harvest fraction corresponding to this maximum is used throughout the model. Then, estimate the maximum harvest rate for the 50-year horizon, if we wish to reduce the extinction probability by one-half.

Exercise 7.7 Multiple Component Failures

Consider a manufacturing process with a number of activities that are arranged in a *serial* and *parallel* configuration. Serial processes mean that if any earlier process fails, the entire process fails, while parallel processes must all fail for the process to fail. This particular manufacturing process has the following chain: A, (B,C,D), E, F, (G,H), I. Letters contained within parentheses are parallel, while the others are serial. Table 7.3 provides data about the mean time between failures and cost for improving the MTBF (used in part [b]) for each of these processes.

TABLE 7.3

Component Failure Rates

Process	MTBF (Hours)	Cost for 10% Increase in MTBF ($Millions)
A	325	16
B	331	21
C	426	14
D	256	10
E	340	20
F	245	18
G	140	14
H	333	15
I	400	12

Assume that each component's failure time can be modeled as an Exponential distribution, with the MTBF as the rate.

a. Build a spreadsheet to simulate the total time to system failure. What is the expected time to failure? Provide a 90% confidence interval.

b. Suppose you can invest $100 million in improving the reliability of these processes. Table 7.2 shows the R&D cost (1 year duration with no uncertainty about success) for achieving a 10% increase in the MTBF for each process. How should you spend the $100 million to achieve the greatest increase in the expected time to system failure (partial projects are not feasible)? How should you invest the $100 million if you want to increase the expected time to system failure as much as possible, while guaranteeing that the 95th percentile of the time to system failure is at least 1100 hours?

Exercise 7.8 Concert Pricing

You are arranging a concert in a venue that seats 1000 people. There are three tiers of seating, as described in Table 7.4:

a. Find the set of prices that maximizes total revenue from this event. Assume that if the seats in a particular tier are not sufficient to satisfy demand, then the unsatisfied fans will not attend at all (i.e., they will not purchase a different tier's tickets—as an optional exercise, you can change this assumption so that unsatisfied demand will switch to the next lower-priced tier if there is room available).

b. Suppose you could reconfigure the venue to permit different proportions of seats in the first two tiers (but the third tier must have 500 seats). What is the revenue maximizing configuration and set of prices?

c. Now suppose demand is uncertain. Table 7.4 provides the mean demand for each tier (using the linear demand as a function of price to provide the mean demand), but each demand is lognormally distributed, with standard deviations of 1000, 80, and 100, respectively (the standard deviations do not depend on the price). What configuration and set of prices maximizes expected total revenue? What is the probability that revenue will not exceed $55,000?

TABLE 7.4

Concert Ticket Tiers

Tier	Seats Available	Demand Function
A	100	1050 − 5 * Price
B	400	20,400 − 250 * Price
C	500	20,000 − 400 * Price

d. Suppose you wish to halve the probability of getting $55,000 (or less) of revenue. Is there a configuration of seats and prices that accomplishes this? If there is, what is the expected total revenue?

Notes

1. See Andrew, B. E. 2007. *The Future of Pricing: How Airline Ticket Pricing Has Inspired a Revolution*. Palgrave Macmillan.
2. Littlewood, K. 1972. "Forecasting and Control of Passenger Booking." *AGIFORS Proceedings, 12 Annual Symposium Proc.*
3. Search conducted on Business Source Premier, March 17, 2019.
4. This is the same hypothetical problem examined in Boyd (2007).
5. We will truncate these distributions to ensure that demand is never negative—since the Normal distribution is continuous, there is a (small) probability of generating a negative demand from these Normal distributions. We could use alternative distributions, but we will stay close to the problem discussed in Boyd (2007).
6. Figure 7.1 shows the demand levels as integers, but in reality, the Normal distribution is producing continuous values. We have chosen to round the demand levels to integer values (both in the display and when calculating sales) to clarify the exposition. Fractional values make sense if we interpret these demand levels as average weekly demands. It would be more appropriate to model demand using a discrete distribution such as the Poisson or Polya distribution, but we are using the same assumptions as Boyd (2007).
7. Strategy 2 generates 21.7% more expected revenue that Strategy 1, while Strategy 3 generates 22.1% more revenue. This compares with Boyd's results of 21.4% and 24.6%, respectively. As we will see, Boyd's application of Littlewood's rule appears to be not exactly accurate.
8. In other contexts, for example, a portfolio choice model, measuring expected return *and risk* may be more important to the investor.
9. Boyd's (2007) use of 37 reserved seats for the application of Littlewood's rule seems to be an error.
10. If you don't see it, it is an add-in that needs to be activated. You can activate it the same way that Data Analysis was activated in Chapter 4.
11. This chapter does not provide a comprehensive overview of different optimization methods and algorithms, and focused on stochastic optimization.
12. In some situations, it can be useful to start with discrete decision variables, even if your decision variable is in reality continuous. It can also be useful to use large steps between discrete values to get a sense of how the objective varies throughout the range of the decision variable. Once you have an idea of where the optimum solution may lie, you can narrow the range, decrease the step sizes, or specify the decision variable as continuous, and repeat the optimization.
13. The capability to set a constraint that seats sold not be greater than the plane's capacity does mean that we could have constructed the model differently, using the RiskOptimizer constraint to handle this. We don't recommend this, however, since in this case it's easy to include the total capacity in the model. It is advisable to use the constraints in RiskOptimizer for constraints that are not easy to accommodate as logic in the model.
14. For larger models, we may even set the number of iterations to 1000 or less. In a second round of optimization, after RiskOptimizer has identified where approximately the optimum may be, we may then run another optimization with a higher number of iterations.
15. In more complex problems, RiskOptimizer may not be able to identify "the absolute, globally best" solution after a limited number of trials or a limited time period. For more complex optimization models with RiskOptimizer, the question is whether RiskOptimizer has found a solution sufficiently close the best one. One way to gauge this is to examine

the top several optimization trials to see if the decision variable(s) that generate these results are reasonably similar or not. If very different sets of decision variables generate similar results for the optimization target, then typically additional iterations should be tried. When the best optimization trials do not entail much variation in the decision variables, then we are more confident that we are close(r) to the optimal solution. Other types of optimization models, such as linear optimization models, do give us a guarantee for a global optimum, but such models come with restrictions on the mathematical form of the model.

16. These assumptions are reflecting that the Y class demand is more inelastic than the Q class demand.
17. The Lognormal distribution could be a better choice for the demand distribution, because there is likely some skew where demand is more often lower than the mean than above it. However, we will continue to use the Normal distribution here to stay consistent with the earlier example.
18. This can be found simply through calculus.
19. This section is more advanced and can be skipped without loss in continuity. Stochastic dominance is not extensively used in management practice, although it is of considerable theoretical interest and is increasingly being employed in some areas. For example, see Sribonchita, S. et al. 2009. *Stochastic Dominance and Application to Finance, Risk, and Economics*. Chapman & Hall/CRC Press.
20. While we don't have an available function to calculate the area under this curve over its entire range, we approximated the area using the calculations shown in Figure 7.27 (a rectangular approximation of the area).
21. A good introduction to QALYs and their measurement can be found in Sassi, F. 2006. "Calculating QALYs, Comparing QALY and DALY Calculations." *Health Policy and Planning*, 21(5):402–408.
22. The data has been adapted from Fotr, J., Plevny, M., Svecova, L. and Vacik, E. 2013. "Multi-Criteria Project Portfolio Optimization under Risk and Specific Limitations." Ekonomika a management.
23. Adapted from Yakubu, A. A., Li, N., Conrad, J. M. and Zeeman, M. L. 2011. "Constant Proportion Harvest Policies: Dynamic Implications in the Pacific Halibut and Atlantic Cod Fisheries." *Mathematical BioSciences*, 232:66–77.

Appendix: Risk Analysis in Projects

Projects are notorious for running over budget and taking more time than planned.[1] When planning for, and executing, projects, it is typical that there are many uncertainties, which make project scheduling and costing an area in which simulation analysis can be very helpful. While many smaller and less complex projects can be analyzed in simple spreadsheets, complex projects are usually planned by using specialized project software. Users of Microsoft Project will find that @Risk is designed to work together to produce probabilistic project schedules and costs. In this appendix, we provide a brief example to illustrate the way @Risk can be used to analyze uncertainty within a project planned in Microsoft Project.

New Product.mpp contains a project plan for a hypothetical new product. The plan is projected to involve 103 tasks, take 154 weeks for completion, and cost $1,213,840. The main steps for conducting a simulation analysis of this project are:

- Import the project into @Risk. This produces a duplicate project in Excel with the same tasks, durations, and costs as in the Microsoft Project file.
- Develop and include the key uncertainties in the @Risk model and set up the simulation to collect the appropriate outputs.
- Run the simulation model in @Risk. This involves @Risk sampling the distributions that were input into the spreadsheet, passing each iteration into Project where the Project plan is updated, and returning the resulting project plan to the spreadsheet.
- When the simulation is complete, the spreadsheet will show the probabilistic outputs for duration and costs.

This process utilizes the capabilities of both @Risk and Microsoft Project, so both programs are required. The first step, importing the project into the spreadsheet, only needs to be done once. After the initial import, the Project file and the @Risk file will automatically be linked—each time the @Risk file is opened, the Project file will then open simultaneously. The Project is imported using "Import mpp file" found under Project on the @Risk toolbar. A progress window will show that all the tasks have been imported, and the spreadsheet will look just like the Project file, including the Gantt chart. There are two other related commands found under Project on the @Risk toolbar.

1. Schedule Audit can be used to ensure that the project is internally consistent and no errors exist in the plan.
2. Sync now is used if you change the project structure (e.g., add or remove tasks rather than just making durations, timing, and/or costs uncertain) to have both the Project and @Risk files consistent.

Once the project is imported (New Product.xlsx contains the imported file), in step #2, uncertainty can now be added to any of the spreadsheet fields. This includes uncertainty about start times, finish times, activity durations, costs, and risk of uncertain events. We will illustrate a number of these features and how they can be included. The simplest way to do this is to simply replace a number with a distribution. For example, the duration for activity 4, "Gather information required for go/no-go decision," has been changed from 1.2 weeks to a Pert distribution (with 1, 1.2, and 1.6 as the minimum, most likely, and maximum parameters, respectively). We have replaced the management cost (on the Resource sheet) of $100/hour with a Pert distribution with minimum, most likely, and maximum values of $90/hour, $100/hour, and $120/hour, respectively. Project costs and durations are often asymmetric, with greater probability of higher rather than lower values.

Many projects are quite large, and entering distributions for individual cells may be unwieldy. Also, many durations and costs may have similar features. For example, all of the tasks in the development stage may be subject to similar uncertainties. Under Project–Model Tools–Risk Categories is an option to create a category of risks with common features. We have chosen all development stage tasks and applied a Pert distribution to the work required (again, asymmetric) for these tasks, as shown in Figure A.1.[2]

Another expedient way to enter several uncertainties is to use another Model Tool, the Parameter Entry Table. Figure A.2 shows an example, where the three engineering resources have been modeled to have the same uncertainty, represented by a Pert distribution (with a minimum value 10% lower than the most likely standard wage rate, and the maximum value 20% higher).

Both of these data entry facilitators permit a variety of choices of what field to apply the uncertainty to, what distribution to use, and what parameters to apply to those distributions.

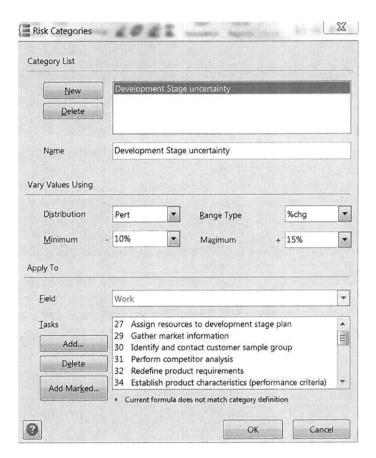

FIGURE A.1
Risk Categories Example

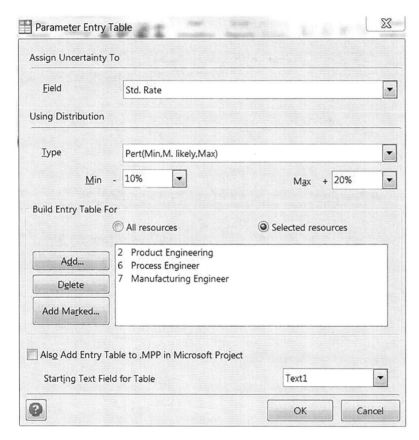

FIGURE A.2
Parameter Entry Table Example

Another common uncertainty entails events that may or may not occur, and that can delay a project. To incorporate this type of event, we create a new worksheet called Risk Register. An example might be that task number 73, "Confirm project safety," may encounter some difficulties that require a more thorough legal investigation of potential liability that might result from a potential safety issue with the new product. We assume that this event has a 30% probability of occurring, and if it transpires, then both the duration and cost may be affected, as shown in Figure A.3.

The key function is found in cell K2, the RiskProjectAddDelay, which instructs @Risk to add the appropriate delay and cost to the relevant task. You must also change the duration and cost of the affected task on the task worksheet—change the original 1 week and $6000 cost to formulas, adding cells J2 and J3 from the Risk Register worksheet. If you recalculate the spreadsheet (looking at the Risk Register worksheet) by clicking F9 a few times (note that this takes more time, since each iteration in the Excel spreadsheet is passed through to Microsoft Project and then returned to Excel), you will see that now and then this episodic risk occurs and some duration and cost is added to the project. When the impact shows TRUE, you will see the delay and added cost reflected on the Task worksheet.

A final modeling tool to note is Probabilistic Branching (found under Project–Model Tools). This permits an event to either delay or accelerate a project. It could be used to delay a project similarly to the safety issue we modeled using the Risk Register. That would require adding a new task to the original project, but a task that may or may not be needed. Rather than modifying the project structure

	A	B	C	D	E	F	G	H	I	J	K
1	Risk	Probability	Simulate	Occurs?	impacts	min	most likely	max	simulated impact	Added	Risk Formula
2	Safety Issue	0.3	=RiskBernoulli(B2)	=IF(C2=1,TRUE,FALSE)	duration impact	2	4	8	=RiskPert(F2,G2,H2)	=$C$2*I2	=RiskProjectAddDelay(TasksIC75,'Risk Register'!J2,'Risk Register'!IJ3,RiskName(A2))
3					cost impact	6000	10000	20000	=RiskPert(F3,G3,H3)	=$C$2*I3	

FIGURE A.3
Risk Register Example

(which can easily be done—but make sure to use Sync to make the project and spreadsheet consistent), we will use the Risk Register to incorporate the potential safety issue. To illustrate the use of probabilistic branching, we will provide an example where there is a chance that the project can be accelerated. Suppose that there is a small chance that the Market Evaluation (tasks 28–31) may be skipped, due to an accelerated timetable if this new product is established as a management priority.[3] Figure A.4 shows how probabilistic branching is set up.

The Add button is used to add the potential tasks that may follow task 26. There is now a 90% chance of following the original schedule (going next to task 28), but a 10% chance of going directly to task 33.

Before running the simulation, you can go to Project Settings and collect Timescaled Data (you can choose what to collect if you want detailed probabilistic output showing monthly cost information, for example). Also, set the total project duration and cost as outputs. We run the simulation to obtain the results, shown in Figure A.5.[4]

One of the useful output reports is a probabilistic Gantt chart. This shows confidence intervals for the durations for the tasks as well as a Critical Index that shows the percent of the simulations where each activity was part of the critical path of the project. Figure A.6 shows part of the display.

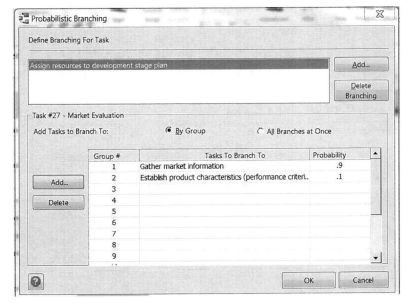

FIGURE A.4
Probabilistic Branching Example

Appendix

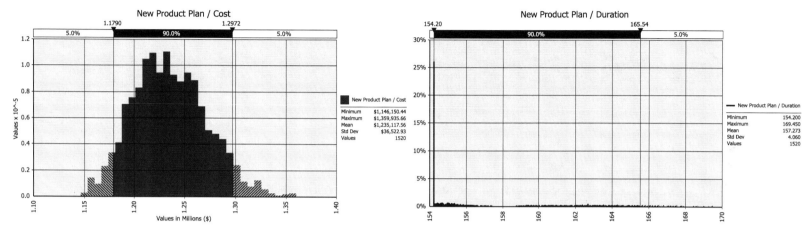

FIGURE A.5
Cost and Duration Results

The probabilistic Gantt chart enables decision makers to focus on those tasks that are most critical to the project duration time. The timescaled data report provides a workbook showing the month-by-month project costs, as in Figure A.7.

We have modified the display in Figure A.7 to make it more legible. The solid black line shows the mean monthly cost from the simulation, and the dashed line shows the deterministic ("base case") costs. The mean simulated costs occur slightly later than in the deterministic case (due to the asymmetries in the project duration where tasks are more likely to take longer times than shorter times). The gray lines indicate the maximum and minimum monthly costs from the simulation. Additional percentiles are provided in the report. Outputs such as these can be used in further modeling of the project, such as the potential profitability of the new product (which would depend on how long it takes to develop the product, as well as its costs).

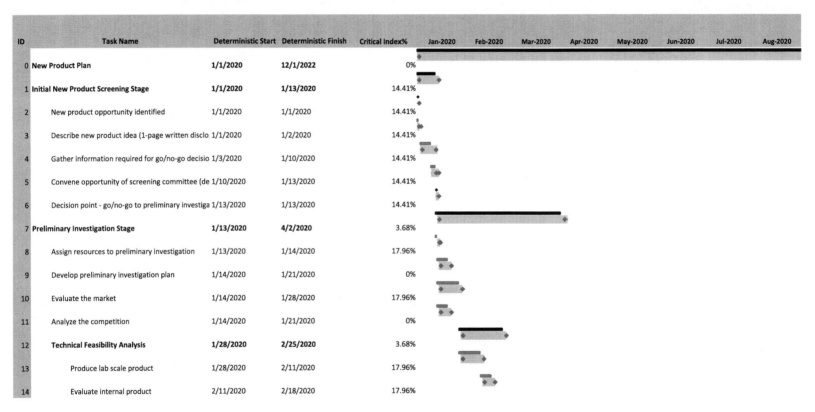

FIGURE A.6
Probabilistic Gantt Chart

Appendix

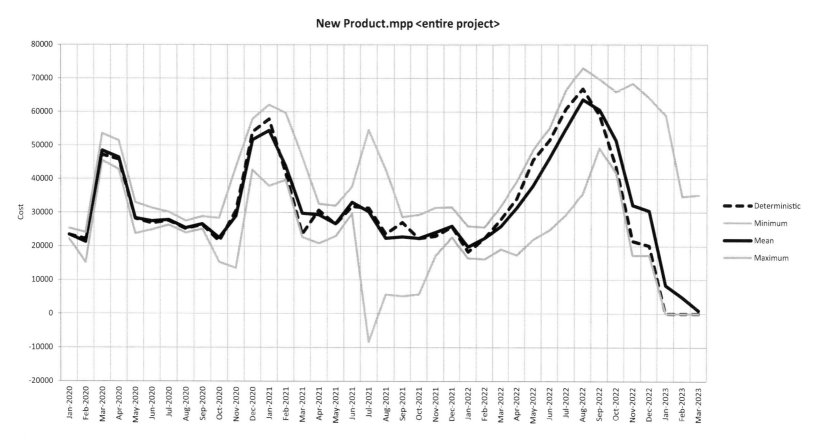

FIGURE A.7
Timescaled Cost Data

Notes

1. For example, Bloch, M., Blumberg, S. and Laartz, J. 2012. "Delivering Large-Scale IT Projects on Time, on Budget, and on Value." *McKinsey Quarterly* claims that large IT projects on average, run 45% over budget, 7% over time, and deliver 56% less value than predicted. Reasons for construction cost overruns are explored in Amusan, L. M., Afolabi, A., Ojelabi, R., Omuh, I. and Okagbue, H. I. "Data Exploration on Factors That Influence Construction Cost and Time Performance on Construction Project Sites." *Data in Brief*, https://doi.org/10.1016/j.dib.2018.02.035, Elsevier, Inc.
2. We recommend that care be taken in eliciting appropriate ranges for each of the probability distributions that represent the various uncertainty in the timeline or costs.
3. We are not necessarily advocating skipping a market evaluation based on management discretion—we only offer this as an example of how probabilistic branching works.
4. The astute reader may notice that we have run fewer simulations than normal. These @Risk/Project simulations run much more slowly than normal, particularly when timescaled data is collected. We still recommend that users run at least 10,000 iterations in their models.

Index

A

Absolute addressing, 11
Advertising effectiveness, 100–105
    Excel regression output, 103
    sales regression line, 102
AIC, *see* Akaiki information criteria
Airline pricing problem
    simulation model of, 177–179
    stochastic optimization solution, 181–187
Airline seat pricing, 177
Air travel and September 11 attacks, 125
    air travel, 127
    Auto Detect, 126
    fit time series for air travel data, 128
Akaiki information criteria (AIC), 58
    goodness-of-fit measure, 127
Amazon (AMZN), 130
American Statistical Association (ASA), 105
Analysis of variance (ANOVA), 103
Anderson-Darling, 58
AND functions, 81
ANOVA, *see* Analysis of variance
Anylogic, 43
Arbitrary judgment, 105
ARCH models, *see* Autoregressive conditional heteroskedasticity models
Arena, 43
ARMA models, *see* Autoregressive moving average models
ASA, *see* American Statistical Association
Autocorrelation, 92
Automated Error Checking, 14
Autoregressive conditional heteroskedasticity models (ARCH models), 133
Autoregressive moving average models (ARMA models), 133

B

Base case scenario, 25
Bayesian analysis, 163
    prior and posterior distributions, 168
    Bayesian model setup, 167
    frequentist confidence/credibility intervals, 167
    rate of claims on insurance premiums, 166
Bayesian approach, 165, 169
Bayesian information criteria (BIC), 58
Bayesian model averaging, 65
Bernoulli distribution, 68–69
"Best"-case scenarios, 25
Beta distribution, 71
BIC, *see* Bayesian information criteria
Blameless errors, 14
Box and whisker plot, 41

C

Capital asset pricing model (CAPM), 49
CDOs, *see* Collateralized Debt Obligations
Claim Frequency distribution, 160
Cockpit failures, 88–92
    cockpit model with correlations, 91
    simulation results, 92
COGS, *see* Cost of goods sold
Collateralized Debt Obligations (CDOs), 84
    model, 87
    results, 88
Communications technologies, 18
Compound distribution, 159–160
    insurance results with, 162
    RiskCompound function, 161
Conceptual maps and models
    decision analysis, 14
    errors in spreadsheet modeling, 12–14
    good practices with spreadsheet model construction, 12
    MoviePass case, 1–2
    retirement planning example, 7–12
    visualization, 2–7
Concert pricing, 205–206
Constrained optimization problems, 190
Constraints, 182
    adding, 190–191

Consumer price index (CPI), 135
Consumer sensitivity, 4
Continuous distributions, 71
Correlation matrix, 88–90
CORREL function, 103
Cost of goods sold (COGS), 50
CPI, *see* Consumer price index
Critical Index, 212
Culpable errors, 14
Cumulative distribution, 73
Customer lifetime value (LTV), 18
    simulation, 45

D

DA, *see* Decision analysis
Data Analysis, 103
Data Grid, 60
Data Validation, 14
Decision analysis (DA), 14
Decision guide for selecting distributions, 73–74
Decision variables, 181, 183, 188
DES, *see* Discrete event simulation
Discrete distributions, 67, 70
    cockpit failure model, 68
    cockpit failure probabilities as function of hours, 69
Discrete event simulation (DES), 42–43
Discrete frequency distribution, 160
Distribution Fitting, 57
Distribution selection
    continuous distributions, 71–73
    decision guide for selecting distributions, 73–74

discrete distributions, 67–71
distribution fitting for value at risk, 56–65
exercises, 74–78
failure distributions, 65–67
valuation model, 50–55
valuation of public company, 49–50
value at risk, 55–56
Drug development, 81–84
    model, 83
    process, 82
    results, 84
    tornado chart for, 85–88

E

Economic order quantity model (EOQ model), 202
Educational investment simulation, 44
Education, investing in, 17–18
Efficient frontier, 191–195
Electric vehicles, 17
    cost simulation, 44
Empirical copulas, 97
    copula fit for three cities' home price changes, 98
    copula to distribution inputs, 97
    hypothetical data, 100
    parametric and empirical copulas, 100
Empirical distributions, 73
eNPV, *see* Expected NPV
Envelope method, 111–113
    model, 113
    trendlines for expert opinion, 112

EOQ model, *see* Economic order quantity model
Errors in spreadsheet modeling, 12–14
European Spreadsheet Risks Interest Group (ESRIG), 13
Evolutionary algorithms, 184–185
Excel, 43
Expected NPV (eNPV), 82
Exponential, distributions, 70, 73

F

Failure distributions, 65
    cockpit failure MTBF rates, 67
Fisheries management, 204
Fitting distributions to data, 55–56, 73
Free cash flows (FCFs), 49
Frequency and severity, 152
    average claim size and expenses per policy, 157
    Bernoulli distribution for claim frequency, 154
    binomial distribution for number of claims, 158
    initial insurance model, 156
    insurance visualization, 153
    lognormal severity distribution, 155
    long and short insurance model results, 159
Fuel costs, 202

G

Gamma distribution, 72
GARCH model, 131–133
Gaussian distribution, 71
GBM models, *see* Geometric Brownian motion models

Index

General distribution, 73, 174
Genetic algorithms, 185
Geometric Brownian motion models (GBM models), 132
Geometric distribution, 70
Goodness of fit (GOF), 65
    statistics, 58

H

Health insurance choices, 17
    simulation, 44
Helios and Matheson Analytics, Inc. (HMNY), 1
Hill-climbing technique, 184
HMNY, see Helios and Matheson Analytics, Inc.
Hollywood finance, 20
    simulation, 45–46
Home prices correlation, 92–97
    copula fit for home price change data, 96
    scatterplot of monthly home price changes, 95
    Seattle and Las Vegas home prices, 93

I

IC, see Information criteria
IF functions, 81
Independent and identically distributed random variables ("iid" random variables), 160
Information criteria (IC), 58
Insurance problem, 151–152
Intentional errors, 14
Inventory position, 202
Investing in education, 17–18

J

Joint likelihood, 58
Jump diffusion, 133

K

Kolmogorov-Smirnov statistics, 58

L

Labor costs, 202
Laplace distribution fits, 58, 61
LINEST function, 108–109, 112
Littlewood's rule, 178–179
Lognormal distribution, 71, 73
Log return, 57
LOOKUP functions, 86, 101
LTV, see Customer lifetime value

M

MA(1) model, see Moving average model
Mean control strategy, 178–179
Mean square (MS), 103
Mean time between failures (MTBF), 65
Microsoft Project, 43
Model uncertainty, 65
Monte Carlo simulation, 2, 27, 185–186
    DES, 42–43
    exercises, 44–46
    models, 169
    Monte Carlo simulation for MoviePass, 28
    random numbers generation, 27–28
    retirement planning, 23
    for retirement planning, 38–42
    using @Risk, 28–38
    risk and uncertainty, 23–25
    Scenario Manager, 25–27
Monte Carlo simulation, 7
MoviePass
    Case, 1–2
    Monte Carlo simulation for, 28
Moving average model (MA(1) model), 125
MS, see Mean square
MTBF, see Mean time between failures
Multiple component failures, 204–205
Multiple decision variables, optimization with, 187–189
Multivariate time series, 134
    five city home price simulation, 137
    model, 135

N

National Household Travel Survey (NHTS), 75
Net present value (NPV), 18
Network economics, 18–19
    simulation, 45
NHTS, see National Household Travel Survey
Noise, 125
Non-parametric copula, 99
Non-parametric distributions, 73, 147
    distribution fit for MPG data, 149
    miles per gallon for hybrid vehicles, 148
Non-parametric fitting, 74
Normal distribution, 71
NPV, see Net present value

O

Objective function, 182
Optimal inventory management, 202–203
Optimization
    adding constraints, 190–191
    airline seat pricing, 177
    efficient frontier, 191–195
    with multiple decision variables, 187–189
    simulation model of airline pricing problem, 177–179
    simulation table to exploring pricing strategies, 179–181
    stochastic dominance, 196–200
    stochastic optimization solution to airline pricing problem, 181–187
OR functions, 81

P

Parameter and model uncertainty, 56–65
    AAPL daily stock price and day-to-day change, 57
    bootstrap option for fitting distributions, 61
    bootstrap parameter confidence intervals, 62
    bootstrap samples, 63, 64
    distributions fit for daily changes in AAPL stock price, 59
    distributions ready for simulation, 60
    simulation results for 5-day AAPL price change, 61
Parameter uncertainty, 60
Parametric bootstrap, 106
Parametric copula patterns, 97
Parametric distributions, 73
Peak load pricing, 19–20
    simulation, 45
Pert distribution, 28, 53
Pessimistic scenario definition, 26
Poisson distribution, 70
Portfolio management, 202
Posterior probability, 164
Pricing decisions, 2
Pricing strategies, simulation table to exploring, 179–181
Probabilistic Branching, 211–212
Probabilistic Gantt chart, 212–214
Probability
    density function, 29
    distribution, 24, 53, 54, 74
    of subscription, 4
    of success, 152
Project management, 20
    simulation, 45
Project planning software, 43
Project selection, 203–204
P-value, 104

Q

Qualitative errors, 14
Quality-adjusted life years (QALYs), 201–202
Quantitative errors, 14

R

Random numbers generation, 27–28
Random walk models, 133
R&D, *see* Research and development
Regression models, 101, 105–107
    bootstrap sample regression, 108
    multiple linear, 107–111
    simulation results, 109
Relative addressing, 11
Reorder point, 202
Research and development (R&D), 83
Retirement planning, 7–12, 23
    Monte Carlo simulation for, 38–42
Revenue passenger mile data (RPM data), 125
Risk analysis in projects, 209
    cost and duration results, 213
    parameter entry table example, 211
    risk categories, 210
    timescaled cost data, 215
RiskCompound function, 160–161
RiskCormat function, 91
RiskDiscrete distribution, 54
RiskOptimizer, 185–186, 188
@Risk packages, 28, 74, 181, 184, 209
    buttons used setting up model, 32
    Copula matrix, 86
    distribution, 28–29
    initial MoviePass model results, 34
    Menu Bar, 30
    Monte Carlo simulation using, 28
    MoviePass model with potential discounted admission prices, 36
    MoviePass model with uncertainty, 31
    MoviePass profit results with reduced admission prices, 37
    MoviePass profit results with reduced consumer sensitivity, 38

permits, 94
simulation output functions in spreadsheet, 35
simulation settings window, 33
toolbar, 147, 153
RiskPercentile function, 34
RiskProjectAddDelay, 211
Risk Register, 211–212
RiskSimTable, 86
RiskTheoPercentile function, 63
RPM data, *see* Revenue passenger mile data

S

Scenario Manager, 25–27
Scenario Summary, 26
Schedule Audit, 209
Second-order distribution fit, 63
Second-order uncertainty, 60
Sensitivity analysis, 143
    tornado graph for advanced, 145
Simple linear regression analysis, 101
Simul8, 43
Simulation model of airline pricing problem, 177–179
Simulation table to exploring pricing strategies, 179–181
Single-leg pricing problem, 177
Solver, 185
Soybean prices, 133
    forecast model, 134
    forecast results, 135
Spreadsheet Engineering Research Project, 13

Spreadsheet model, 2
    errors in, 12–14
    good practices with construction, 12
    retirement, 9
Standard Error, 104
Statistical theory, 104
STEYX function, 106
Stochastic dominance, 196–200
Stock prices, 130
    static and time series models for stock price data, 132
    stock price simulation model, 132
    ten years of daily stock prices and daily price changes, 131
Stress testing, 145
    spider graph, 146
    stress analysis box plots, 147
    stress testing setup, 146
Student-t distribution, 71
Sync now, 209

T

Time series models, 132–133
    air traffic data and 9/11 attacks, 125–130
    dialog, 123–124
    multivariate time series, 134–137
    soybean prices, 133–134
    stock prices, 130–132
    types, 132–133
Time series patterns, 57
Trace Dependents, 13
Trace Precedents, 13
Triangular distribution, 53, 73

Time series analysis, 121
    distribution of time series, 122
    forecast for detrended data, 124
    forecast with deseasonalization, 125
    hypothetical time series data, 122

U

Uncertainty, 2, 23–25, 163
    realistic insurance model, 163
    retirement planning visualization with, 39
    and variability, 161
Uniform distribution, 73

V

Value at Risk (VaR)
Valuation model, 50
    Facebook simulation results, 55
    initial Facebook, 51–52
    inputs to valuation model, 53
Valuation of public company, 49
    Facebook Income Statements, 50
VisiCalc (spreadsheet program), 12–13
Visualization, 2–7
    retirement, 8
VLOOKUP functions, 86

W

Weibull distribution, 73
Weighted average cost of capital (WACC), 49
"Worst"-case scenarios, 25